THE LORE OF
THE HONEY-BEE

By

TICKNER EDWARDES

First published in 1908

Copyright © 2019 Home Farm Books

This edition is published by Home Farm Books,
an imprint of Read Books Ltd.

This book is copyright and may not be reproduced or copied in any
way without the express permission of the publisher in writing.

British Library Cataloguing-in-Publication Data
A catalogue record for this book is available
from the British Library.

www.readandcobooks.co.uk

To
The Chairman of
The British Bee-Keepers Association,

THOMAS WILLIAM COWAN

To whose labours and researches
the writer, and all other Bee-Keepers,
are under a lasting obligation.

CONTENTS

Bee Keeping 7

INTRODUCTION
THE OLDEST CRAFT UNDER THE SUN 11

CHAPTER I
THE ANCIENTS AND THE HONEY-BEE 21

CHAPTER II
THE ISLE OF HONEY 33

CHAPTER III
BEE-MASTERS IN THE MIDDLE AGES 40

CHAPTER IV
AT THE CITY GATES 57

CHAPTER V
THE COMMONWEALTH OF THE HIVE 71

CHAPTER VI
EARLY WORK IN THE BEE-CITY 83

CHAPTER VII
THE GENESIS OF THE QUEEN 90

CHAPTER VIII
THE BRIDE-WIDOW 107

CHAPTER IX
THE SOVEREIGN WORKER-BEE 114

CHAPTER X
A ROMANCE OF ANATOMY 128

CHAPTER XI
THE MYSTERY OF THE SWARM 147

CHAPTER XII
THE COMB-BUILDERS 164

CHAPTER XIII
WHERE THE BEE SUCKS 182

CHAPTER XIV
THE DRONE AND HIS STORY......................192

CHAPTER XV
AFTER THE FEAST................................202

CHAPTER XVI
THE MODERN BEE-FARM 208

CHAPTER XVII
BEE-KEEPING AND THE SIMPLE LIFE 216

ILLUSTRATIONS

The Comb-Builders, With Chain
of Wax-Making Bees 22

Moses Rusden's Bee-Book 43

A Page from Butler's "Bees' Madrigall," 1623 48

Rev. John Thorley Writing his "Melissologia"
with the Help of his Bees. 1744
(From an old Bee Book) 53

Inverted Straw Skep-Hive, Showing
Natural Arrangement of Combs 61

An Old Sussex Bee-House 65

The City Gate 71

Comb-Frame from Modern Hive, with Queen 77

Drone-Brood and Worker-Brood 93

Queen Bee in Breeding-Season. 103

A Queen-Cell 105

The Honey-Bee (Enlarged) From Life: and as
Some of the Ancient Draughtsmen Depicted her 117

Brood-Comb, Showing the Two
Sizes of Cell Together, with Eggs
Cemented by Queen to Cell-Bases 126

The Bee-Nursery: Tending the Young Brood 144

A Swarm in May 150

A Mammoth Swarm 154

Hiving The Swarm. 157

The Swarm Hived..................................162

Honey-Comb by Transmitted Light,
Showing Arrangement of Cells on Both Sides..........175

Brood-Comb Showing Eggs, Larvæ in
Different Stages of Growth, Sealed Cells,
and Young Bees Cutting Their Way Out...............182

In the Store-House: Sealing up the New Honey.........192

Chapman Honey-Plant in Village Garden.............211

Bad Beemanship: Bees Lying-Out from a
Too Crowded Hive..................................216

A Forest Apiary......................................222

Bee Keeping

Beekeeping (or apiculture, from Latin: apis 'bee') is quite simply, the maintenance of honey bee colonies. A beekeeper (or apiarist) keeps bees in order to collect their honey and other products that the hive produces (including beeswax, propolis, pollen, and royal jelly), to pollinate crops, or to produce bees for sale to other beekeepers. A location where bees are kept is called an apiary or 'bee yard.' Depictions of humans collecting honey from wild bees date to 15,000 years ago, and efforts to domesticate them are shown in Egyptian art around 4,500 years ago. Simple hives and smoke were used and honey was stored in jars, some of which were found in the tombs of pharaohs such as Tutankhamun.

The beginnings of 'bee domestication' are uncertain, however early evidence points to the use of hives made of hollow logs, wooden boxes, pottery vessels and woven straw baskets. On the walls of the sun temple of Nyuserre Ini (an ancient Egyptian Pharo) from the Fifth Dynasty, 2422 BCE, workers are depicted blowing smoke into hives as they are removing honeycombs. Inscriptions detailing the production of honey have also been found on the tomb of Pabasa (an Egyptian nobleman) from the Twenty-sixth Dynasty (c. 650 BCE), depicting pouring honey in jars and cylindrical hives. Amazingly though, archaeological finds relating to beekeeping have been discovered at Rehov, a Bronze and Iron Age archaeological site in the Jordan Valley, Israel. Thirty intact hives, made of straw and unbaked clay, were discovered in the ruins of the city, dating from about 900 BCE. The hives were found in orderly rows, three high, in a manner that could have accommodated around 100 hives, held more than 1 million bees and had a potential annual yield of 500 kilograms of honey and 70 kilograms of beeswax!

It wasn't until the eighteenth century that European understanding of the colonies and biology of bees allowed the construction of the moveable comb hive so that honey could be harvested without destroying the entire colony. In this 'Enlightenment' period, natural philosophers undertook the scientific study of bee colonies and began to understand the complex and hidden world of bee biology. Preeminent among these scientific pioneers were Swammerdam, René Antoine Ferchault de Réaumur, Charles Bonnet and the Swiss scientist Francois Huber. Huber was the most prolific however, regarded as 'the father of modern bee science', and was the first man to prove by observation and experiment that queens are physically inseminated by drones outside the confines of hives, usually a great distance away. Huber built improved glass-walled observation hives and sectional hives that could be opened like the leaves of a book. This allowed inspecting individual wax combs and greatly improved direct observation of hive activity. Although he went blind before he was twenty, Huber employed a secretary, Francois Burnens, to make daily observations, conduct careful experiments, and keep accurate notes for more than twenty years.

Early forms of honey collecting entailed the destruction of the entire colony when the honey was harvested. The wild hive was crudely broken into, using smoke to suppress the bees, the honeycombs were torn out and smashed up — along with the eggs, larvae and honey they contained. The liquid honey from the destroyed brood nest was strained through a sieve or basket. This was destructive and unhygienic, but for hunter-gatherer societies this did not matter, since the honey was generally consumed immediately and there were always more wild colonies to exploit. It took until the nineteenth century to revolutionise this aspect of beekeeping practice – when the American, Lorenzo Lorraine Langstroth made practical use of Huber's earlier discovery that there was a specific spatial measurement between the wax combs, later called the bee space, which bees do not block with wax, but keep as a free passage. Having determined this bee space (between 5 and 8 mm, or 1/4 to 3/8"), Langstroth then designed a series of wooden frames within a rectangular hive box, carefully maintaining the correct space between successive frames, and found that the bees would build parallel honeycombs in the box without bonding them to each other or to the hive walls.

Modern day beekeeping has remained relatively unchanged. In terms of keeping practice, the first line of protection and care – is always sound knowledge. Beekeepers are usually well versed in the relevant information; biology, behaviour, nutrition - and also wear protective clothing. Novice beekeepers commonly wear gloves and a hooded suit or hat and veil, but some experienced beekeepers elect not to use gloves because they inhibit delicate manipulations. The face and neck are the most important areas to protect (as a sting here will lead to much more pain and swelling than a sting elsewhere), so most beekeepers wear at least a veil. As an interesting note, protective clothing is generally white, and of a smooth material. This is because it provides the maximum differentiation from the colony's natural predators (bears, skunks, etc.), which tend to be dark-coloured and furry. Most beekeepers also use a 'smoker'—a device designed to generate smoke from the incomplete combustion of various fuels. Smoke calms bees; it initiates a feeding response in anticipation of possible hive abandonment due to fire. Smoke also masks alarm pheromones released by guard bees or when bees are squashed in an inspection. The ensuing confusion creates an opportunity for the beekeeper to open the hive and work without triggering a defensive reaction.

Such practices are generally associated with rural locations, and traditional farming endeavours. However, more recently, urban beekeeping has emerged; an attempt to revert to a less industrialized way of obtaining honey by utilizing small-scale colonies that pollinate urban gardens. Urban apiculture has undergone a renaissance in the first decade of the twenty-first century, and urban beekeeping is seen by many as a growing trend; it has recently been legalized in cities where it was previously banned. Paris, Berlin, London, Tokyo, Melbourne and Washington DC are among beekeeping cities. Some have found that 'city bees' are actually healthier than 'rural bees' because there are fewer pesticides and greater biodiversity. Urban bees may fail to find forage, however, and homeowners can use their landscapes to help feed local bee populations by planting flowers that provide nectar and pollen. As is evident from this short introduction, 'Bee-Keeping' is an incredibly ancient practice. We hope the current reader is inspired by this book to be more 'bee aware', whether that's via planting appropriate flowers, keeping bees or merely appreciating! Enjoy.

INTRODUCTION

THE OLDEST CRAFT UNDER THE SUN

ONE of the oldest and prettiest fables in ancient mythology is that which deals with the origin of the honey-bee. It was to Melissa and her sister Amalthea, the beautiful daughters of the King of Crete, that the god Jupiter was entrusted by his mother Ops, when Saturn, his father—following his custom of devouring his children at birth—sought to make the usual meal of this, his latest offspring.

The story is variously rendered by ancient writers. Some say that bees already existed in the world, and that Amalthea was only a goat, whose milk served to nourish the baby-god, in addition to the honey that Melissa obtained from the wild bees in the cave where Jupiter lay hidden. Another account has it that the bees themselves were drawn to his place of concealment by the noise made by his nurses, who beat continually on brazen pans to keep the sound of his infant lamentations from the ears of his ravening sire. Thenceforward the bees took over the charge of him, bringing him daily rations of honey until he grew up and was able to hold his own in the Olympian theogony. In either case Jupiter showed his gratitude towards his preservers in true celestial fashion. It was a very ancient belief among the earliest writers that, in the single instance of the honey-bee, the ordinary male-and-female principle was abrogated, and that the propagation of the species took place by miraculous means. In explanation of this, we are told it was a special gift from Jupiter in acknowledgment of the unique service rendered him. In one version of the fable, and in the words of a famous bee-master who wrote in 1657, "Jupiter, for so great a benefit, bestowed on his nurses for a reward that they should have young ones, and continue their kind, without wasting themselves in venery." In the other, and probably much older form of the legend, Melissa, the beautiful Princess of Crete, was herself changed by the god into a bee, with the like immaculate propensities; and thenceforward the work of collecting honey for the food of man—that honey which, down to a very few centuries from the present time, was universally believed to be a miraculous secretion from heaven—was confided to her descendants.

Apart, however, from the old dim tales of ancient mythology, where there is a romance to account for all beginnings of the world and everything upon it, any attempt to trace back the art of bee-keeping to its earliest inception cannot fail to bring us to the conclusion that it is inevitably and literally the oldest craft under the sun. Thousands of years before the Great Pyramid was built, bee-keeping must have been an established and traditional occupation of man. It must have been common knowledge, stamped with the authority of the ages, that a bee-hive, besides its toiling multitudes, contained a single large ruling bee, divine examplar of royalty; for how else would the bee have been chosen to represent a King in the Egyptian hieroglyphic symbols? But it is not only within the limit of historical times, however remote, that evidences of bee-culture, or at least of man's use of honey and wax in his daily life, are to be found or inferred. So far back as the Bronze Age it is certain that wax was used in casting ornaments and weapons. A model of the implement was first made in some material that would perish under heat. This was imbedded in clay, and the model burnt out, after which the mould thus formed was filled with the molten metal. These models, no doubt, were in many cases carved out of wood; but it is certain that another and more ductile material was often used. Bronze ornaments have been found with thumb-marks upon them, obviously chance impressions on the original model faithfully reproduced. And the substance of these models could hardly have been anything else than beeswax.

But speculation on the probable antiquity of bee-keeping need not stop here. The best authorities estimate that human life has existed on the earth for perhaps a hundred thousand years. The earliest traces of man, far back in the twilight of palæolithic times, reveal him as a hunting and fighting animal, in whom the instinct to cultivate the soil or domesticate the creatures about him had not yet developed. Later on in the Stone Age—but still in infinitely remote times—it is evident that he tamed several creatures, such as the ox, the sheep, and the goat, keeping them in confinement, and killing them for food as he required it, instead of resorting to the old ceaseless roaming after wild game. At this time, too, he took to sowing corn, and even baking or charring some sort of bread. It must be remembered that if a hundred thousand years is to be set down as the limit of man's life on the earth, probably the development of other living creatures, as well as most forms of vegetable life, took place immeasurably earlier. The chances are that the world of trees and flowering-plants, in which aboriginal man moved, differed in no great degree from the world of green things surrounding human life today. It is certain that the apple, pear, raspberry, blackberry, and plum were common fruits of the country-side in the later Stone Age, for seeds of all these have been found in conjunction with neolithic remains. Evidence of the existence of the beech and elm—the latter a famous pollen-yielder—has been discovered at a very much earlier time. All the conditions favourable to insect-life must have been present in the world ages before man appeared in it; and insect-life undoubtedly existed then in a high state of development. It would be as unreasonable, therefore, not to infer that the honey-bee was ready on the earth with her stores of sweet-food for man, as that man did not speedily discover that store, and make it an object of his daily search, just as he went forth daily to hunt and kill four-footed game.

There is, of course, a great deal of difference between a chance discovery of a wild-bee's nest, as a common and expected incident in a day's foraging, and the systematic preservation and tending of bee-hives as a source of daily food. While it is reasonable to assume that the first men used honey as an article of diet, it is probable that they were a wandering race, never halting for long in the same locality, and therefore unlikely to be bee-keepers in the accepted sense of the word. They depended, no doubt, on the wild honey-stores which they happened to find in their entourage for the time being. But the first sign of civilisation must have been the gradual lessening of this nomadic instinct. Tribes would come to take permanent possession of districts rich in the game, as well as the fruits and tubers, necessary for their daily food. At the same time the haunts of the wild bees would be discovered, their enemies kept down or driven away, the places where the swarms pitched annually noted, and thus the first apiary would have been founded, probably long before any attempt at cultivation of the soil or domestication of the wild creatures for food was made.

Biologists generally regard hunting as the oldest human enterprise under the sun; but, adopting their well-known method of deductive reasoning, it seems possible to make out a rather better case for beemanship in this category. The primæval huntsman must have found much difficulty in bringing down his game, and still more in securing it, when maimed, but yet capable of eluding final capture. For this purpose some sort of retrieving animal, fleeter of foot and more cunning than its master, must have been even more necessary in primæval times than it is in the modern days of the gun. There seems to be no evidence of man indicating the most elementary civilisation without sure signs also that he had trained and used some sort of dog to help him in his daily food-forays. But man must have existed long before civilisation can be said to have come within age-long distance of him. In these times, beset with enemies, he must have built his hut nest-like in some high, impregnable tree, out of reach of night-prowling foes; and it is scarcely conceivable that the dog was his companion under these conditions. More probably he lived, for the most part, on fruits and honey-comb, and such of the small creatures as he could capture with his naked hands. Thus, in all likelihood, the first hunter was a bee-hunter. Eolithic man may have had his own rocky fastness or clump of hollow trees, where the wild bees congregated; and with the coming of each summer he may have followed his swarms through the glades of primæval forests as zealously as any bee-keeper of the present day.

Speculation of this kind is necessarily far-fetched and fantastic, and can be but half seriously undertaken with so small and inconsiderable a creature as the honey-bee. But it is interesting from one special, and not often adopted, point of view. There is no more fascinating study than that of the ancient civilisations of the world. Egypt 10,000 years ago, Babylon probably still earlier, China that seems to have stopped at finite perfection in all ways that matter little, ages before the time of Abraham. But all these are of mushroom growth compared with the antiquity of bee-civilisation. It is only a tale of Lilliput, of a microscopic people living and moving on a mimic stage. Yet, perhaps tens of thousands of years before man had made fire, or chipped a flint into an axe-head, these winged nations had evolved a perfect plan of life, and solved social problems such as are only just beginning to cloud the horizon of human existence in the twentieth century. And they, and their intricate communal polity, have not passed away into dust, as the great human nations of bygone ages have done, and as those of the present day may be destined to do, for all we can tell.

Will a time come when we must learn from the honey-bee or perish? We have still probably a few thousand years wherein to think it out, and prepare for it; but unless the world comes to an end, or human increasing-and-multiplying comes to an end, one earth will eventually become too small to hold us. With this thought in mind, a study of the honey-bee and the arrangements of hive-life, takes on a new interest. Supposing that the political economy of a bee-hive may be taken as a foreshadowing of the ultimate human state, there is no denying that we get a glimpse into an eminently disquieting state of things, at least from the masculine point of view. We see matriarchy triumphant; the females holding supreme control in the State, and not only initiating all rules of public conduct, but designing and carrying through all public works. The male is reduced to the one indispensable office of sex, and even a single exercise of this is vouchsafed only to a few in a thousand. But to create the large and permanent army of workers necessary in a State such as this, and to recruit it wholly from the females, it became necessary to revise all rules of life from their very foundation. There must have been a great renunciation among the bees, male and female alike, when the resolve was made to leave the whole duty of procreation of their kind to one pair alone of their number—one pair only out of every thirty thousand or so—in order that the rest could devote themselves to ceaseless, sexually uncommoded toil.

This may be imagined as following on a great discovery, an epoch-making discovery, changing the whole face and future of bee-life—how, by the nursing and feeding of the young grub of the female bee, she could be atrophied into a mere, sexless, over-intellectual labourer, or glorified into a creature lacking, it is true, all initiative and almost all mental power, but possessing a body capable of mothering the whole nation. Here is socialistic political economy carried to its sternest, most logical conclusions. All is sacrificed for the good of the State. The individual is nothing: the race is everything. "Thorough" is the motto of the honey-bee, and she drives every theory home to its last notch. Men are pleased to call themselves bee-masters; but the best of them can do no more than study the ways of their bees, learn in what directions it is their will to move, and then try to smooth the way for them. The worker-bees collectively are the whole brains in the business, and the bee-keeper is as much the slave of the conditions and systems they have inaugurated as they are themselves; while the queen-bee is the most willing, and, at certain seasons, the most laborious slave of them all.

It is useless to deny that bee-polity, with its stern dead-reckoning of ingenuity, its merciless adherence to the demands of a system perfected through countless ages, has its unpleasant and even its revolting aspects. Nature is always wonderful, but not always admirable; and a close study of the Life within the Hive brings out this truth perhaps more clearly than with any other form of life, humanity not excepted. Absolute communism implies incidental cruelty: it is only under a system of bland political compromise, of neighbourly give and take, that justice and mercy can ever be yoke-fellows. In the republic of bees, nothing is allowed to persist that is harmful or useless to the general good. Every individual in the hive seems to acquiesce in this common principle—either by choice or compulsion—from the mother-bee down to the last lazy drone, born into the brief plenty of waning summer days. In the height of the honey-flow, the State demands a storehouse filled to the brim; and every bee keeps herself to the task unceasingly until death from overwork comes upon her, and her last load never reaches the hive. If the queen-bee grows old, or her powers of egg-laying prematurely fail, she is ruthlessly slaughtered, and her place filled by another specially raised by the workers to meet this contingency during her lifetime and in her full view. Drones are bred in plenty, plied with the richest provender in the hive, and allowed to wanton through their days of insatiate appetite, so that no young queen may go forth on her nuptial flight unchallenged. But when the last princess is happily mated, and safely home again in the warm, awaiting cluster, every drone is callously done to death, or driven out of the hive to perish. If hard times threaten, or the supply of stores is arrested, the old and worn-out members of the hive are exterminated, breeding is stopped, the unborn young are torn from their cradle-cells and destroyed, so that there may be as few mouths as possible to fill in the lean days to come. The signs of dawning prosperity or adversity are watched for, and the working population of the hive is either increased or checked, just as future probabilities seem to indicate.

But the most bewildering, most uncanny thing of all about this bee-republic is the fact that, in it, has been successfully solved the problem of the balance of the sexes. While all other creatures in the universe bring forth their kind, male and female, in what seems a haphazard, unpremeditated way, these mysterious hive-people cause their queen-mother to give them either sons or daughters according to the needs of the community. They lead her to the drone-cells, and she forthwith deposits eggs that hatch out infallibly as drones; and in the combs specially made wherein to rear the aborted females, the workers, the queen is caused to lay eggs that just as assuredly produce only the worker-bee.

It is the oldest civilisation in the world, this wonderful commonwealth of the bee-people, and it is not unprofitable to examine it in the light of ideas which are at present only flickering up uncertainly on the distant path, but which might well broaden out some day into general conflagration. It is conceivable that a time existed when the conditions of bee-life were very different from those we see to-day. Bees have drawn together into vast communities, just as men are slowly, but surely, gathering into cities. A time may come when individual existence outside the city may be as impracticable for men, as life has become for separate bee-families away from the hive; and then there may arise a purely masculine dilemma. It may be that once the magnificent drone was of real consequence in domestic affairs. Bee-life may have consisted of numberless small families, each with its deep-voiced, ponderous father-bee, its fruitful mother, and its tribe of youngsters growing up, and in time setting forth to establish homes for themselves. There is no reason why each one of the thirty or forty thousand pinched virgins in a hive should not have become a fully developed, prolific queen-bee, if only the right food, in sufficient quantity, had been given her in her larval state. But the need for the single large community arose. The system of a single national mother was instituted. The great renunciation was made, for good or ill. And then the trouble, from the masculine point of view, began.

It must be borne in mind that, strictly speaking, the honey-bee does not, and never did, possess a sting. What is commonly known as her sting is really an ovipositor, and it is as such that it is almost exclusively used by the modern queen-bee in every hive to-day. But when the first hordes of worker-bees were brought into the world, reduced by the science of starvation to little more than sexless sinews and brains, they seemed to have conceived a terrible revenge on their ancestors. The useless ovipositor was turned into a weapon of offence, against which the drone's magnificent panoply of sound and fury availed him nothing. Matriarchy was established at the point of the living sword. A pitiless logic overran everything. Intolerance of all the bright asides of life—the wine, the dance, the merry talk, and genial tarryings by the path, beloved of all drones, bee or human—darkened the day. And the result is only more honey, a vaster storehouse filled to the brim with never-to-be-tasted sweets, at a cost unfathomable, when the old larder would have sufficed for every real need, and life might still have been merry and leisurely.

It is only a fable, far-fetched, fantastic, as any told to the Caliph in the "Arabian Nights." But there, again, the woman had her way, like the bee-woman before, and some day she and her kind may get it on a more ambitious scale. And then—what of the sword that was once a sewing-needle?

> "Some are content with saying that they do it by Instinct, and let it drop there; but I believe God has given us something farther to do, than to invent names for things, and then let them drop."
>
> —A. I. Root.

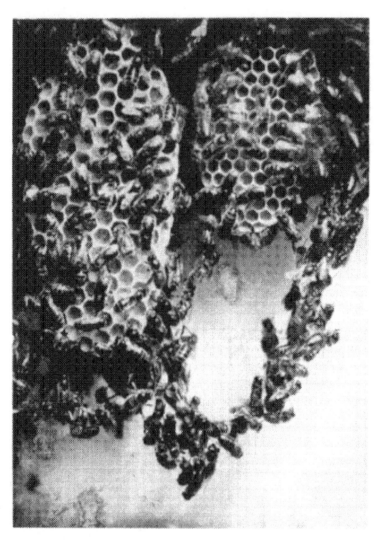

*The Comb-Builders,
With Chain of Wax-Making Bees*

THE LORE OF THE HONEY-BEE

CHAPTER I

THE ANCIENTS AND THE HONEY-BEE

"While great Cæsar hurled War's lightnings by high Euphrates, ... even in that season I, Virgil, nurtured in sweet Parthenope, went in the ways of lowly Quiet."

—*Fourth Book of the Georgics.*

IT was in Naples—the Parthenope of the Ancients—that the "best poem by the best poet" was written, nearly two thousand years ago. Essentially an apostle of the Simple Life, the cultured and courtly Virgil chose to live a quiet rural existence among his lemon-groves and his bee-hives, when he might have dwelt in the very focus of honour at the Roman capital; where his friend and patron, Mæcenas, the prime minister of Octavian, kept open house for all the great in literature and art.

Modern bee-keepers, athirst for the Americanisation of everything, give little heed nowadays to the writings of one whom Bacon has called "the chastest poet and royalest that to the memory of man is known." And yet, if the question were asked,

What book should first be placed in the hands of the beginner in apiculture to-day? no wiser choice than this fourth book of the Georgics could be made.

For Virgil goes direct to the great heart of the matter, which is the same to-day as it was two thousand years ago. The bee-keeper must be first of all a bee-lover, or he will never succeed; and Virgil's love for his bees shines through his book from beginning to end. Of course, in a writer so deeply under the spell of Grecian influences, it is to be expected that such a work would faithfully reproduce most of the errors immortalised by Aristotle some three hundred years before. But these only serve to bring the real value of the book into stronger relief. Through the rich incrustation of poetic fancy, and the fragrant mythological garniture, we cannot fail to see the true bee-lover writing directly out of his own knowledge, gathered at first hand among his own bees.

Virgil knew, and lovingly recorded, all that eyes and ears could tell him about bee-life; and it is only within the last two hundred years or so that any new fact has been added to Virgil's store. All the writers on apiculture, from the earliest times down to the eighteenth century, have done little else than pass from hand to hand the fantastic errors of the ancient "bee-fathers," adding generally still more fantastic speculations of their own. And until Schirach got together his little band of patient investigators of hive-life about a hundred years ago, Virgil's fourth Georgic—considered as a practical guide to bee-keeping—was still very nearly as well-informed and up-to-date as any.

It is not, however, for its technical worth that the book is to be recommended to the apiarian tiro of to-day. All that has become hopelessly old-fashioned with the passing of the ancient straw-skep in the last generation. The intrinsic value of Virgil's writings lies in their atmosphere of poetry and romance, which ought to be held inseparable, now as ever, from a craft which is probably the most ancient in the world. Almost alone among country occupations to-day, bee-keeping can retain much of its entrancing old-world flavour, and yet live and thrive. But if the modern tendency to make the usual unlovely transatlantic thing of British honey-farming is to be checked, nothing will do more to that end than an early instillation of Virgil's beautiful philosophy.

Dipping into this fascinating poem—with its delightful blend of carefully told fact, and rich fancy, and quaint garnerings from records then extant, but now lost in the ages—we can reconstruct for ourselves a picture of Virgil's country retreat near "sweet Parthenope," where he loitered, and mused, and wrought the faultless hexameters of the Georgics with so much care and labour, that the work took seven years to accomplish—which is at the rate of less than a line a day.

Virgil's house stood, probably, on the wooded slope above the town of Naples, deep set in orange-groves and lemon-plantations, and in full view, to the north, of the snow-pinnacled Apennines, and, southward, of the blue waters of the Bay. Vesuvius, too, with its eternal menace of grey smoke, rose dark against the morning sun only a few leagues onward; and, at its foot, the doomed cities nestled, Pompeii and Herculaneum, then with still a hundred years of busy life to run.

Bee-hives in Virgil's day—as we can gather from certain ancient Roman bas-reliefs still in existence—were of a high, peaked, dome pattern, and they were made of stitched bark, or wattled osiers, as he himself tells us. Many of the directions he gives as to their situation and surroundings are still golden rules for every bee-keeper. The bee-garden, he says, must be sheltered from winds, and placed where neither sheep nor butting kids may trample down the flowers. Trees must be near for their cool shade, and to serve as resting-places when "the new-crowned kings lead out their earliest swarms in the sweet spring-time." He tells us to place our hives near to water, or where a light rivulet speeds through the grass; and we are to cast into the water "large pebbles and willow-branches laid cross-wise, that the bees, when drinking, may have bridges to stand on, and spread their wings to the summer sun."

Virgil's method of hiving a swarm is almost identical with that followed by old-fashioned bee-men to this day. The hive is to be scoured with crushed balm and honeywort, and then you are to "make a tinkling round about, and clash the cymbals of the Mother"—that is, of the goddess Cybele. The bees will forthwith descend, he tells us, and occupy the prepared nest. When the honey-harvest is taken, you are first to sprinkle your garments and cleanse your breath with pure water, and then to approach the hives "holding forth pursuing smoke in your hand." And the old-time bee-man of to-day takes his mug of small-beer as a necessary rite, and washes himself before handling his hives.

But perhaps the great charm of the fourth Georgic consists, not in its nearness to truth about bee-life, but in the continual reference to the beautiful myths, and hardly less attractive errors, of immemorial times, copied so faithfully by mediæval writers, but not apt to be heard of by the learner of to-day unless he reads the old books.

Virgil begins his poem by speaking of "heaven-born honey, the gift of air," in allusion to the belief that the nectar in flowers was not a secretion of the plant itself, but fell like manna from the skies. He seriously warns his readers of the disastrous effect of echoes on the denizens of a hive, and of the hurtful nature of burnt crab-shells; and tells us that in windy weather bees will carry about little pebbles as counterpoises, "as ships take in sand-ballast when they roll deep in the tossing surge."

He was a firm believer in the Divine origin of bees. To all the ancients the honey-bee was a perpetual miracle, as much a sign and token of an omnipotent Will, set in the flowery meadows, as is the rainbow, to modern pietists, set in the sky. While all other creatures in the universe were seen to produce their kind by coition of the sexes, these mysterious winged people seemed to be exempt from the common law. Virgil, copying from much older writers, says, "they neither rejoice in bodily union, nor waste themselves in love's languors, nor bring forth their young by pain of birth; but alone from the leaves and sweet-scented herbage they gather their children in their mouths, thus sustaining their strength of tiny citizens."

Just as marvellous, however—at least to the modern entomologist—will appear the belief, widespread among the ancients, and shared by Virgil, that swarms of bees can be spontaneously generated from the decaying carcass of an ox. Virgil professes to derive his account of the matter from an old Egyptian legend, and he gives careful directions to bee-keepers of what he seems never to doubt is an excellent method for stocking an apiary. There is a very old translation of the passage in the fourth book of the Georgics relating to these self-generated bees, which is worth quoting, if only on account of its quaint mediæval savour. "First, there is found a place, small and narrowed for the very use, shut in by a leetle tiled roof and closed walles, through which the light comes in askant through four windowes, facing the four pointes of the compass. Next is found a two-year-old bull-calf, whose crooked horns bee just beginning to bud; the beast his nose-holes and breathing are stopped, in spite of his much kicking; and after he hath been thumped to death, his entrails, bruised as they bee, melt inside his entire skinne. This done, he is left in the place afore-prepared, and under his sides are put bitts of boughes, and thyme, and fresh-plucked rosemarie. And all this doethe take place at the season when the zephyrs are first curling the waters, before the meades bee ruddy with their spring-tide colours, and before the swallow, that leetle chatterer, doethe hang her nest again the beam. In time, the warm humour beginneth to ferment inside the soft bones of the carcase; and wonderful to tell, there appear creatures, footless at first, but which soon getting unto themselves winges, mingle together and buzz about, joying more and more in their airy life. At last, burst they forth, thick as rain-droppes from a summer cloude, thick as arrowes, the which leave the clanging stringes when the nimble Parthians make their first battel onset."

For a study in the persistence of delusions, this affords us some very promising material. In the first place, the generation of bees from putrescent matter is, and must always have been, an impossibility. If there is one thing that the honey-bee abhors more than another, it is carrion of any description. Indeed, putrid odours will often induce a stock of bees to forsake its hive altogether; so it cannot even be supposed that bees would venture near the scene of Virgil's malodorous experiment, and thus give rise to the belief that they were nurtured there. But not only was this practice a recognised and established thing in Virgil's time, but entire credence was placed in it throughout the Middle Ages down, in fact, to so late a time as the seventeenth century. It is on record that the experiment was carried through with complete success by a certain Mr. Carew, of Anthony, in Cornwall, at an even later date still.

The practice, moreover, was of infinitely greater antiquity than even Virgil supposed. He was probably right in giving it an Egyptian origin, and this alone may date it back thousands of years. In Egypt the custom had a curious variant. The ox was placed underground, with its horns above the surface of the soil. Then, when the process of generation was presumed to be complete, the tips of the horns were sawn off, and the bees are said to have issued from them, as out of two funnels.

Nearly all the ancient writers, with the exception of Aristotle, mention the practice in some form or other. Varro, writing half a century before Virgil, says, "it is from rotten oxen that are born the sweet bees, the mothers of honey." Ovid gives the story of the Egyptian shepherd Aristaeus as enlarged upon by Virgil, and adds some speculations of his own. He suggests that the soul of the ox is converted into numberless bee-souls as a punishment to the ox for his lifelong depredations amongst the flowers and herbage, the bee being a creature that can only do good to, and cannot injure, vegetation.

Manifestly, where there is so general, and so widely independent a corroboration of a story, some explanation must exist, which will alike bear out the truth and condone, or at least extenuate, the error. A careful examination of the various accounts of bee-swarms having been produced from decaying animal matter reveals one common omission in regard to them. All the writers are agreed that dense clouds of bee-like insects are evolved; and speak of these as escaping into the air and flying off, presumably in the immediate quest of honey. But no one bears testimony to honey having been actually gathered by these insects, nor is it recorded that they were ever induced to take possession of a hive, as ordinary swarms of bees will readily do. They are spoken of more as enriching the neighbourhood generally, by augmenting the number of bees abroad, than as conducing to the well-being of any particular bee-owner.

Herein, no doubt, is to be found a clue to the whole mystery. If it was not the honey-bee—the *Apis mellifica* of modern naturalists—which was generated from the entombed body of Virgil's unfortunate bull-calf, what other insect, closely resembling a bee, could have been produced under those conditions? The answer has been readily given by several naturalists of our own time. There is a fly, called the drone-fly, which exactly meets the difficulty. He is so like the ordinary honey-bee that on one occasion, and that recently, he was mistaken for the genuine insect by one calling himself a bee-expert, and holding a diploma officially entitling him to the use of that name. This drone-fly would have behaved almost exactly as Virgil's calf-bred bees are said to have behaved, and according to the various descriptions of the matter given by other writers living before and since. He would issue forth in a dense cloud immediately his natal prison-doors were opened, and he would comport himself in other ways exactly as enumerated. Finally, he would beget himself joyously to the open country, as a swarm of bees would do; and once more the Virgilian theory of bee-production would meet with its seeming verification.

But having gone thus far with the drone-fly, it is difficult to resist going a little farther. We cannot leave him in the ignominious company of slaughtered oxen, but must give him his due of more lordly associations. "Out of the eater came forth meat, and out of the strong came forth sweetness." When Samson went down to Timnath on his fateful mission of wooing, and saw the carcass by the way beset with a cloud of insects, we need not cast any doubt on his genuine belief that they were honey-bees. He propounded his riddle in all good faith, and the form of it can very well be explained as a not undue stretch of allowable poetic privilege. But that the creatures he saw hovering about the dead lion were really bees, and that Samson actually obtained honey from the carcass, is not to be accepted without the exercise of a faith that is undistinguishable from credulity. Many attempts have been made to explain away the difficulties of the problem on natural lines, but they are all alike unconvincing. There is little doubt at this time that the part of the story dealing with the honey is nothing but a deft embroidering on the original legend by some later chronicler; and that the insects which were seen about the dead lion were really drone-flies generated in the same fashion as those from Virgil's ox.

Perhaps no better general idea is to be obtained of the condition of bee-knowledge among the ancients than from the writings of Pliny, the Elder, who was born in A.D. 23. He, too, deals with the ox-born bees; but the reader's interest will centre for the most part in Pliny's grave and careful account of the life and customs of the honey-bee, as commonly accepted among his contemporaries. Very few indeed of the facts he so picturesquely details have any real foundation in truth. Like nearly all the classic writers, he had little more accurate knowledge of the life within the hive than we have of the bottom of the Pacific Ocean. But he made up for this deficiency, as did all others of his time, by dipping largely into the stores of his own fancy as well as those of other people.

His account of the origin and nature of honey is quaintly pleasant reading. "Honey," he says, "is engendered from the air, mostly at the rising of the constellations, and more especially when Sirius is shining; never, however, before the rising of the

Vergiliæ, and then just before daybreak.... Whether it is that this liquid is the sweat of the heavens, or whether a saliva emanating from the stars, or a juice exuding from the air while purifying itself—would that it had been, when it comes to us, pure, limpid, and genuine, as it was when first it took its downward descent. But, as it is, falling from so vast a height, attracting corruption in its passage, and tainted by the exhalations of the earth as it meets them; sucked, too, as it is, from off the trees and the herbage of the fields, and accumulated in the stomachs of the bees, for they cast it up again through the mouth; deteriorated besides by the juices of flowers, and then steeped within the hives and subjected to such repeated changes:—still, in spite of all this, it affords us by its flavour a most exquisite pleasure, the result, no doubt, of its æthereal nature and origin."

Modern bee-keepers ascribe the varying quality in honey nowadays to the prevalence of good or bad nectar-producing crops during the time of its gathering, or to its admixture with that bane of the apiculturist—the detestable honey-dew. But Pliny set this down entirely to the influence of the stars. When certain constellations were in the ascendant, bad honey resulted, because their exudations were inferior. Honey collected after the rising of Sirius—the famous honey-star of all the ancient writers—was invariably of good quality. But when Sirius ruled the skies in conjunction with the rising of Venus, Jupiter, or Mercury, honey was not honey at all, but a sort of heavenly nostrum or medicament, which not only had the power to cure diseases of the eyes and bowels, and ameliorate ulcers, but actually could restore the dead to life. Similar virtues were possessed by honey gathered after the appearance of a rainbow, provided—as Pliny is careful to warn us—that no rain intervenes between the rainbow and the time of the bees' foraging.

On the life-history of the honey-bee Pliny wrote voluminously. He tells us of a nation of industrious creatures ruled over by a king, distinguished by a white spot on his forehead like a diadem. These king-bees were of three sorts—red, black, and mottled;

but the red were superior to all the rest. He appears to accept, though guardedly, the old legend that sexual intercourse among bees was divinely abrogated in favour of a system of procreation originating in the flowers. He mentions a current belief—which must have been the boldest of heresies at the time—that the king-bee is the only male, all the rest being females. The existence of the drones he explains away very ingeniously. "They would seem," he says, "to be a kind of imperfect bee, formed the very last of all; the expiring effort, as it were, of worn-out and exhausted old age, a late and tardy offspring."

The discipline in the hives was, according to Pliny, a very rigid affair. Early in the morning the whole population was awakened by one bee sounding a clarion. The day's work was carried through on strict military lines, and at evening the king's bugler was again to be observed flying about the hive, uttering the same shrill fanfaronade by which the colony was roused at daybreak. After this note was heard, all work ceased for the day, and the hive became immediately silent.

His book abounds in curious details as to hivelife. When foraging bees are overtaken in their expeditions by nightfall, they place themselves on their backs on the ground, to protect their wings from the dew, thus lying and watching until the first sign of dawn, when they return to the colony. At swarming-time, the king-bee does not fly, but is carried out by his attendants. Pliny warns intending bee-keepers not to place their hives within sound of an echo, this being very injurious to the bees; but, he adds, the clapping of hands and tinkling of brass afford bees especial delight. He ascribes to them an astonishing longevity, some living as long as seven years. But the hives must be placed out of the reach of frogs, who, it seems, were fond of breathing into hives, this causing great mortality among its occupants. When bees need artificial food, they are to be supplied with raisins or dried figs beaten to a pulp, carded wool steeped in wine, hydromel, or the raw flesh of poultry. Wax, Pliny says, is best clarified by first boiling it in sea-water, and then drying it in the light of the moon, for whiteness. And in taking honey from the hives, a person must be well washed and clean. Malefactors are cautioned against approaching a hive of bees at any time. Bees, he assures us, have a particular aversion to a thief.

To the latter-day practical bee-keeper, all these minute details given by the classic writers read very like useless and cumbersome nonsense; and it seems matter for wonder that the bees contrived to exist at all under such ingeniously complicated mismanagement, born, as it was, of an ignorance flawed by scarcely a single ascertained fact. But the truth stands out pretty clearly that bee-keeping two thousand years ago was really a very large and important industry. One apiary is mentioned by Varro as yielding five thousand pounds of honey yearly, while the annual produce of another brought in a sum of ten thousand sesterces. Pliny mentions the islands of Crete and Cyprus, and the coast-country of Africa, as producing honey in great abundance. Sicily was famous for the good quality of its beeswax, but Corsica seems to have been one of the main sources of this. When the island was subject to the Romans, it is said that a tribute of two hundred thousand pounds' weight of wax was yearly exacted from it. This, however, is such an astounding figure that it must be taken with a certain caution.

Evidently the bees in the ancient world managed their business in fairly good fashion, in spite of the ignorance of their masters, or at least of the ancient chroniclers *de re rustica*. But it should always be borne in mind that the writers on husbandry and kindred subjects were seldom practical men. With the single exception, perhaps, of Virgil's "Georgicon," these old books relating to apiculture bear unmistakable evidence of being, for the most part, merely compilations from writings still more ancient, or heterogeneous gatherings together of hearsays and current fables of the time. It is certain that the men who were actually engaged in the craft of bee-keeping, and who knew most about it, wrote nothing at all. Probably they concerned themselves very little with the myths and fables of bee-craft, and owed their success to hard, practical, everyday experience, which is the surest, and perhaps the only, guide to-day.

CHAPTER II

THE ISLE OF HONEY

IF we are to accept all that the old Roman historians have put on record to the glory of their race, we must believe that their conquering legions found everywhere barbarism, and left in its place the seeds of a high civilisation—high, at least, in the general acceptance of the word in those lurid, moving days.

But it may well be questioned whether the Britain that Cæsar first knew was as barbaric as it has been painted. We are accustomed to look upon Cæsar's account of his earliest view of Albion—of Eilanban, the White Island, as the Britons themselves called it—as the first glance vouchsafed to us into the history of our own land. But this is very far from being the truth. British history begins with the record of the first voyage of the Phœnicians, who, adventuring farther than any other of their intrepid race, chanced upon the Scilly Isles and the neighbouring coast of Cornwall, and thence brought back their first cargo of tin.

And how long ago this is who shall say? The whereabouts of the Phœnician Barat-Anac, the Country of Tin, remained a secret probably for ages, jealously guarded by these ancient mariners, the first true seamen that the world had ever known. They were expert navigators, venturing enormous distances oversea, even in King Solomon's time; and that was a thousand years before the advent of Cæsar. In all likelihood they had been in frequent communication with the Britons centuries before the Greeks took to searching for this wonderful tin-bearing land, and still longer before the name Barat-Anac became corrupted into the

Britannia of the Romans. And it is hardly to be supposed that a people of so ancient a civilisation, and of so great a repute in the sciences and refinements of life, as the Phœnicians—a people from whom the early Greeks themselves had learned the art and practice of letters—could remain in touch, century after century, with a nation like the Britons without affecting in them enormous improvement and development in every way that would appeal to so high-mettled and competent a race.

For high-mettled and capable the Britons were even in those old, dim, far-off days. Cæsar's account of them, read between the lines, accords ill with the commonly accepted notion of a horde of savages, pigging together in reed hovels, and daubing their naked bodies blue to strike terror into the equally savage minds of their island adversaries. We get a glimpse of a people much farther advanced in the arts of peace and war. In all probability they clothed themselves at ordinary times, picturesquely enough, in the furs of the wild animals, with which the island abounded; and it was only in war-time that they stripped and painted. Old prints have familiarised us with the sight of the sailors of Drake and Nelson stripped much in the same way; and the blue paint of Druidical times is not divided by so great a gulf as the ages warrant from the scarlet cloth and glittering brass-ware of nineteenth-century fighting-men. As armourers the ancient Britons must have been not immeasurably inferior to the Romans, and we are told that they excelled in at least one difficult craft, the making of all sorts of basket-ware.

But there is other testimony, apart from Cæsar's, in favour of the view that they were by no means a barbarous people. Diodorus Siculus, who was Cæsar's contemporary, speaks of them as possessing an integrity of character even superior to that commonly obtaining among the Romans; and Tacitus, writing about a century later, ascribes to them great alertness of apprehension, as well as high mental capacity. Protected as they were by the sea, it is probable that war entered to no large extent into their lives, and they were essentially a pastoral people.

The cultured and daring Phœnician traders are certain to have prospected the coast much farther eastward than is recorded, and thus to have materially hastened British advance in civilisation—at least, as far as the southern tribes were concerned.

It has been claimed—on what evidence it is difficult to determine—that the Romans, besides teaching the Britons all other arts of manufacture and husbandry, introduced the practice of bee-culture into the conquered isles. But Pliny, giving an account of the voyages of Pytheas, which are supposed to have been undertaken some three hundred years before Cæsar ever set foot here, mentions the Geographer of Marseilles as landing in Britain, and finding the people brewing a drink from wheat and honey. There is, however, another source of testimony on this point, of infinitely greater antiquity than any yet enumerated. Long before the Phœnician sailors discovered their tin-country, there were bards in Eilenban—the White Island—hymning the prowess of their Celtic heroes and the traditional doings of their race. These old wild songs were handed down from singer to singer through the ages, and many of them, still extant among the records of the Welsh bards, must be of unfathomable antiquity. These profess to describe the state of Britain from the very earliest beginnings of the human race. And in some of them, which are seemingly among the oldest, Britain is called the Isle of Honey, because of the abundance of wild bees everywhere in the primæval woods. There would be little profit, and no little folly, in seeking to invest these old traditions with any more than their due significance. But there is much in a name. And it may be conjectured that if Britain was known among the early Druidical bards as the Isle of Honey the natural conditions giving rise to the name were still prevalent, and reflected immemorially in the life of the people, when Cæsar first saw them crowding the white cliffs above him, a huge-limbed, ruddy-locked, war-like race. He records that they possessed their herds of tame cattle and their cultivated fields; and it is reasonable to suppose that the hives of wattled osier that Virgil wrote of a century later had their ancient counterpart of woven basket hives in the British villages of the day.

No doubt the Romans, during their second and permanent occupation, which did not take place until a hundred years after, taught the Britons their own methods of bee-management, and improved in numberless ways on the practice of the craft, which, among the British, was probably a very simple and rough-and-ready affair. But it was not until the Romans had gone, and the Anglo-Saxon rule was fairly established in the Island, that bee-keeping seems to have become one of the recognised national industries. The records bearing on the social life of the people at that time are necessarily broken and scanty; but it is certain that honey, with its products, had become an important article of diet among all classes, high and low. It is difficult—here in the present time, when cane and beet-sugar, and even chemical sweetening agents, are in constant and universal use—to realise that, from the remotest times down to the fifteenth and sixteenth centuries, there was practically no other sweet-food of any description, except honey, in the world; and to estimate, therefore, what a prominent place in the industries of each country bee-keeping must then have occupied. There was nothing else but honey for all purposes, and it is constantly mentioned in the old monkish chronicles and the curious manuscript cookery-books that have survived from the Middle Ages.

It is true that the sugar-cane was known as far back as the first century A.D. Strabo, writing just before the commencement of the Christian Era, relates how Nearchus, who was Admiral of the Fleet to Alexander the Great, made an important voyage of discovery in the Indian Ocean, and brought back news of the wonderful "honey-bearing reed," which he found in use among the natives of India. There is a record that the Spaniards brought the sugar-cane from the East, and planted it in Madeira early in the fifteenth century. Thence its cultivation spread to the West Indies and South America, during that and the following century. Throughout the Middle Ages it was in very restricted use among the richest and noblest families in Europe, Venice being then the centre of its distribution. But cane-sugar was little else than a costly luxury of diet, or a vehicle in medicine, even among the highest in the land, until well into the seventeenth century, when it slowly began to oust honey from the popular favour. The chances are, however, that the middle and lower classes of England possessed, and could afford, no other sweetening agent but honey, for any purpose, down to about three hundred years ago.

Among the Anglo-Saxons the bee-hives supplied the whole nation, from the King down to the poorest serf, not only with an important part of their food, but with drink and light as well. We read of mead being served at all the royal banquets, and in common use in every monastery. Even in those far-off days there were wayside taverns where drink was retailed; and the chief potion was mead, although a kind of ale was also brewed. No priest was allowed to enter these hostelries, but this could scarcely have been a great deprivation, as the home allowance of mead was a sufficiently generous one. Ethelwold's allowance to each half-dozen of his monks at dinner was a sex-tarium of mead, which, in modern measure, would be probably several gallons.

There were three kinds of liquor brewed from honey in Anglo-Saxon times. The commonest, or mead proper, which may be taken as the usual drink of the masses, was made by steeping in water the crushed refuse of the combs after the honey had been pressed from them. This would be strained and set aside in earthen vessels until it fermented and became mead. And the longer it was kept, the more potent grew the liquor. Another kind, made from honey, water, and the juice of mulberries, was called Morat; and this, presumably, was the beverage of the more well-to-do. A third concoction, known as Pigment, was brewed from the purest honey, flavoured with spices of different sorts, and received an additional lacing of some kind of wine. Probably this was the mead served at the royal table. The office of King's

Cup-bearer could have been no sinecure in those days, for it was the custom of Anglo-Saxon monarchs to entertain their courtiers at four banquets daily, and the quantities of liquor which the old records tell us were consumed on these occasions seem incredible, even in the annals of such a deep-drinking race. Not the least valuable outcome of the Norman Conquest, as far as the national temperance was concerned, must have been the reform instituted in these Court orgies by William the First, who reduced their number to a single state banquet daily.

If it may be supposed that the reign of Harold marked the summit of popularity for our good old English honey-brew, it is equally certain that with the coming of the Normans began its slow decline in the national estimation. Following in the trail of Duke William's nondescript army came the traders, with their outlandish liquors from the grape; and wine must soon have taken the place of the Saxon mead, first among the foreign nobles, and later among the native thanes. From that day mead has steadily declined in vogue, and today mead-making is practically a lost art, surviving only among a few old-fashioned folk here and there in remote country places.

But it is still to be obtained; and those of us who have had the good fortune to taste good old mead, well matured in the wood, are sure to feel a regret that no determined effort is being made to rehabilitate it in the national favour. Perhaps there is no more wholesome drink in the world, and certainly none requiring less technical skill in the making. All the ancient books on bee-keeping give receipts for its manufacture, differing only in the variety of foreign ingredients added for its improvement, or, as we prefer to believe, to its degradation. For the finest mead can be brewed from pure honey and water alone, and any addition of spices or other matter serves only to destroy its unique flavour. Some of the sixteenth and seventeenth century bee-masters were renowned in their day for their mead-brewing; and one of the foremost of them claims for his potion that it was absolutely indistinguishable, by the most competent judges, from old

Canary Sack. He gives careful directions for the manufacture of his mead; and these can be, and have, indeed, recently been, followed with complete success. This mead, when kept for a number of years, froths into the glass like champagne, but stills at once, leaving the glass lined with sparkling air-bells. It is of a pale golden colour, and has a bouquet something like old cider; but its flavour is hardly to be compared with any known liquor of the present time. It is interesting, however, to have its originator's authority for its close resemblance to Canary Sack, as this gives a clue to the intrinsic qualities of a wine long since passed out of the popular ken.

CHAPTER III

BEE-MASTERS IN THE MIDDLE AGES

STUDENTS of old books on the honey-bee are generally struck with two very remarkable characteristics about them—their invariable fine old classic and romantic flavour, and their ingenious leavening of a great mass of quite obvious fable by a very small modicum of enduring fact.

It is difficult to realise, until one has delved deep into these curious old records, how completely they are dyed through and through with the picturesque, but mainly erroneous, ideas of the ancient classic bee-fathers. The writers were, almost without exception, earnest, practical men, whose chief interest in life was the study and pursuit of their craft. But they seem, one and all, to have laboured under the idea that it was their bounden duty to uphold everything written about bees by the old Greek and Roman *litterati*, and that it would be the rankest heresy to advance any new truth, garnered from their individual experience, unless it could be supported by ample testimony from the same infallible source.

They seemed to look upon the works of Aristotle, Virgil, Pliny, and the rest, as so many divine revelations of the mystery of bee-craft, all-sufficing, finitely perfect; and they continually quoted from them in support of their own contentions, or in refutation of the statements of others, much as teachers of religion refer doubters to Bible texts. The bee-masters of the Middle Ages were, however, not alone in adopting this peculiar attitude of mind.

Moses Rusden's Bee-Book

It seems to have been the prevailing habit of the time with all classes. One might almost be justified in concluding that the study of nature in those days had no other object with these inveterate old classicians but to support what had already been set down by their revered oracles. It was enough that a thing had been written in Greek or Latin in the literary youth of the world; it was immaculate — the first and last word on the question; and if their personal observations seemed at variance with any statement of the old-world writers, then the contradiction was only an apparent one, and could, no doubt, be easily resolved by a more learned exponent of these bee-scriptures of ancient days.

It is certainly, at first glance, a matter for wonder that men could pass their whole lives in the pursuit of the craft, and yet manage to preserve uncorrupted a faith which seems so readily, and at so many points, assailable. But it must be remembered that any observation of the inner life of the honey-bee was then an extremely difficult thing. It was next to impossible to see anything that was going on inside the hives in use at that day. Pliny mentions a hive made of what he calls mirror-stone, which was probably talc, and through the transparent sides of which the working of the bees could be seen. But nothing of the kind seems to have been attempted among English bee-masters until the seventeenth century. Moreover, even if the whole hive had been made of clear glass, the observer would have been very little the wiser. He would have had the outer sides of the two end combs in view, and he would have seen much coming and going among the bees, with an occasional glimpse of the queen. But all the wonderful activity of the hive, so laboriously ascertained by latter-day observers, with the help of so many ingenious appliances, goes on entirely in the hidden recesses of the combs; and any attempt to study this life under the conditions appertaining in the Middle Ages would have been manifestly futile. It was not until Huber's leaf-hive was invented—when it became to some extent possible to divide the combs for a short time without hopelessly disturbing the bees—that any real progress in bee-knowledge was made. The modern observation-hive, wherein the bees are compelled to build their combs between glass partitions, one over the other instead of side by side, was a still greater advance, and rendered the whole interior of the bee-dwelling available for study. But it is open to objection that bee-life in such a contrivance is carried on under too artificial conditions. In a natural bee-nest, the combs are built roughly side by side, and the brood is reared in the centre area of each comb, the surface covered by the breeding-cells diminishing outwards in each direction. Thus the brood-nest takes a globular form, with the honey-stores above and around it; and this natural arrangement is inevitably destroyed in a hive where the combs are superimposed and not collateral.

In the face, therefore, of the practical impossibility of learning anything about bees when they were housed in the usual straw-skep, the old bee-masters confined themselves to a repetition of the beliefs of the ancient writers, deftly interwoven with speculations of their own, which, as no one was in a position to refute them, were advanced with all the more daring and assurance.

They seem to have been, in the main, agreed on the point that the ordinary generative principle, otherwise universal throughout creation, was miraculously dispensed with in the single case of the honey-bee. Moses Rusden, who was bee-master to King Charles the Second, and who published his "Further Discovery of Bees" so late as the year 1679, believed that the worker-bees gathered from the flowers not only the germs of life, but the actual corporeal substance, of the young bees.

He pointed triumphantly to the little globular lumps of many-coloured pollen which bees so industriously fetch into the hives during the breeding-season, and asserted that these were the actual bodily matter from which the young bees developed. He also maintained that every hive was ruled over by a king, but here Rusden was evidently trying to serve two masters. No doubt he was a true "Abhorrer," and heartily detested anything at variance with the doctrine of the divine right of monarchs. He had faithfully copied from Virgil as to the gathering of this generative substance from the flowers; but he felt that, as the King's Bee-Master, it was incumbent on him to put in a good word for the restored monarchy if he could. There were still many in the realm who were altogether opposed to the Restoration, and probably more who were waverers between the faiths. And Rusden, doubtless, saw that if he could point to any parallel instance in Nature where the system of monarchy was the divinely ordained state, he would be furnishing his patron with a magnificent argument in favour of his kingship, and one, moreover, which would especially appeal to the ignorant and superstitious masses. No doubt, however, in taking up this position, Rusden was only echoing the belief immemorially established among the bee-men of the past.

The single large bee, which all knew to exist in each hive, was generally looked upon as the absolute ruler of the community. It is variously described as a king or queen by writers in the sixteenth and seventeenth century, but only in the sense of a governor; and the word chosen largely depended on the sex of the august person who happened to occupy the English throne at the time. Thus Rusden very wisely discarded the notion of a queen-bee when he had to deal with Charles the Second. Butler, perhaps the most learned of the mediæval writers on the honey-bee, as astutely forbore to mention the word king, his book being published in the reign of Queen Anne. He calls it "The Feminine Monarchie," but seems to have no more suspected the truth that the large bee was really the mother of the whole colony than any of his predecessors. Almost alone in his day, however, he refuses to accept the flower theory of bee-generation, and asserts that the worker-bees and drones are the females and males respectively. But, he says, they "engender not as other living creatures; onely they suffer their Drones among them for a season, by whose Masculine virtue they strangely conceive and breed for the preservation of their sweet kinde." He gets over the difficulty of there being no drones in the hive for nine months in the year, during part of which time breeding goes actively forward, by asserting that the worker-bees immaculately conceive of the drones for the season, their summer impregnation sufficing until the drones reappear in the May of the following year. Thus, without guessing it, he was very near the discovery of one of the most astounding facts in Nature—that the queen-bee of a hive, after a single traffic with a drone, continues to produce fertile eggs for the rest of her life, which may extend to as long as three, or even four, years.

Butler's book is rich in the quaint bee-lore of his times. He tells us the queen-bee has under her "subordinate Gouvernours and Leaders. For difference from the rest they beare for their crest a tuft or tossel, in some coloured yellow, in some murrey, in manner of a plume; whereof some turne downward like an

Ostrich-feather, others stand upright like a Hern-top. In less than a quarter of an hour," he assures us, "you may see three or foure of them come forth of a good stall; but chiefly in Gemini, before their continuall labour have worne these ornaments." And any warm spring or summer morning, if you watch a hive of bees at work, you may chance upon much the same thing. In some flowers, notably the evening primrose, the pollen-grains have a way of clinging together in threads; and these festoons often catch in the antennæ of the foraging bees, giving much the same appearance of a plume, or tassel, as Butler saw in his day. He gives some advice as to the deportment of a good bee-master which is well worth quoting. "If thou wilt have the favour of thy Bees that they sting thee not, thou must avoid such things as offend them: thou must not be unchaste or un-cleanely: for impurity and sluttishnesse (themselves being most chaste and neat) they utterly abhor: thou must not come among them smelling of sweat, or having a stinking breath, caused either through eating of Leekes, Onions, Garleeke, and the like; or by any other meanes: the noi-somenesse whereof is corrected with a cup of Beere: and therefore it is not good to come among them before you have drunke: thou must not be given to surfeiting and drunkennesse: thou must not come puffing and blowing unto them, neither hastily stir among them, nor violently defend thy selfe when they seeme to threaten thee; but softly moving thy hand before thy face, gently putting them by: and lastly, thou must be no Stranger unto them. In a word, thou must be chaste, cleanly, sweet, sober, quiet, and familiar: so will they love thee, and know thee from all other."

Thus, the good bee-master, according to Butler, is necessarily a compendium of all the virtues; and nothing more seems to be wanted to bring about the millennium than to induce all mankind to become keepers of bees.

Writers on the honey-bee in mediæval times vied with each other in their testimony to the extraordinary powers and intelligence of their hive-people.

A Page from Butler's "Bees' Madrigall," 1623

But perhaps a story, gravely related by Butler, outdoes them all. He prefaces it by declaring that "Bees are so wise and skilful, as not onely to discrie a certaine little God amightie, though he came among them in the likenesse of a Wafer-cake; but also to build him an artificial chappell." He goes on to relate that "a certaine simple woman, having some stals of Bees that yeelded not unto hir hir desired profit, but did consume and die of the murraine; made hir mone to an other Woman more simple than hir selfe; who gave her counsell to get a consecrated Host, and put it among them. According to whose advice she went to the priest to receive the host: which when she had done, she kept it in hir mouth, and being come home againe she took it out, and put it into one of hir hives. Whereupon the murraine ceased, and the Honie abounded. The Woman, therefore, lifting up the Hive at the due time to take out the Honie, saw there (most strange to be seene) a Chappell built by the Bees, with an altar in it, the wals adorned by marvellous skill of Architecture, with windowes conveniently set in their places: also a doore and a steeple with bells. And the Host being laid upon the altar, the Bees making a sweet noise, flew around it."

This story is only paralleled by another, equally ancient, wherein it is related that some thieves broke into a church, and stole the silver casket in which the holy wafers were kept. They found one wafer in the box, and this they hid under a hive before making off with the more intrinsically valuable part of their booty. In the night, it seems, the owner of the hive was awakened by the most ravishing strains of music, coming at set intervals from the direction of his bee-garden. He went out with a lantern to ascertain the cause of it, and discovered it to proceed from the interior of one of his hives. Full of perturbation at this miracle, he went and roused the Bishop, and acquainted him with the extraordinary state of affairs; and the Bishop coming with his retinue and lifting up the hive, they found that the bees had taken possession of the consecrated wafer, and placed it in the upper part of their hive, having first made for it a box of the whitest wax, an exact replica of the one stolen. And all around this box there were choirs of bees singing, and keeping watch over it, as monks do in their chapel. "With which story," adds the narrator prophetically, "I doubt not but some incredulous people will quarrell."

In their directions for hiving a swarm, the mediæval bee-masters were always quaintly explicit. The dressing of the skep which was to receive the swarm was a particularly elaborate process. When the skep was new, you were recommended to scour it out with a handful of sweet herbs, such as thyme, marjoram, or hyssop; and this was to be followed by a second dressing of honey and water, or milk and salt. But the preparation of an old skep must have been a rather disgusting affair. You were to put "two or three handfuls of mault, or pease, or other corne in the hive, and let a Hogge eat thereof. Meanwhile, doe you so turne the Hive, that the fome or froth, which the Hogge maketh in eating, may goe all about the Hive. And then wipe the Hive lightlie with a linnen cloth, and so will the Bees like this Hive better than the new."

When the swarm was up, and "busie in their dance," you were to "play them a fit of mirth on a Bason, Warming-pan, or Kettle, to make them more speedily light." We are assured that the swarm would fly faster, or slower, according to the noise made. If the fit of mirth were in rapid measure, the bees would fly fast and high; but with a soft leisurely music, they would go slowly, and soon descend. This curious custom of "ringing the bees" is undoubtedly of Roman origin; but whether it was introduced by Cæsar's followers, or those of Claudius in the first century, or whether the old English bee-masters themselves derived it from their classic reading, is hard to determine. It is still to be heard in many country districts, and its exponents seem to retain all the faith of their forefathers in its efficacy. Probably, in mediæval times, when bee-gardens were much more plentiful than they are now, the custom had at least one undeniable merit: it proclaimed to the various hive-owners in the vicinity that a swarm was in the air, and that its rightful owner was on the alert. In this way, no doubt, dishonest claims to its possession were largely prevented, or, at least, discouraged.

The question whether the noise made by ringing has any real effect on the swarming bees is still not absolutely decided. With the exception of the old skeppists, not a few of whom still exist in out-of-the-way rural corners, modern apiculturists have long discarded the custom as a gross superstition. But it has recently been suggested that the din made by old-fashioned bee-keepers when a swarm is up may have a real use after all. It is conjectured that the cloud of bees—which at first is nothing but a chaos of flashing wings, the whole contingent darting and whirling about indiscriminately over a large area together—is really dispersing in search of the queen. The suggestion put forward is that they follow her by ear, as she is supposed to utter a peculiar piping sound when flying. The din of the key and pan may, it is said, prevent the bees hearing this note and following her in her first erratic convolutions, and thus the swarm is more likely to pitch on a station near home. The theory is interesting, but hardly tenable. Old popular observances of this kind are seldom based on even the vaguest thread of fact, and it is much more probable that no effect whatever is produced on the bees by the ringing.

With regard to the right of a bee-keeper to follow his swarm into a neighbour's land, it is interesting to have the assurance of one of these ancient writers that "if they will not be stayed, but, hasting on still, goe beyond your bounds; the ancient Law of Christendome permitteth you to pursue them whithersoever, for the recovery of your owne." But, the writer adds, if your swarm goes so fast and so far that you lose sight and hearing of them, you also lose all right and property in them. In this case you have no legal alternative but to leave the bees to whomsoever may first find them. In view of recent disputes on this matter, wherein the law laid down appears to have been both vague and arbitrary, it is useful to be able to point to so ancient an authority in vindication of the bee-keeper's rights.

There is hardly any detail in bee-government which had not its curious observance or superstition in mediæval times. One and all seemed to believe in the old Virgilian notion that bees carried about little stones to balance their flight during windy weather, and some even thought that flowers were carried about in the same way. Red-coloured clothing was supposed to be particularly offensive to bees, and one is warned not to venture near the apiary thus attired. In the hives the old bees and the young were believed to occupy separate quarters. In regard to this, it is a well-attested fact that, during the height of the honey season, the bees found in the upper stories of a hive are principally young ones who have not yet flown.

We are told that if any of the bees have not returned to the hive at the end of the day, the queen goes out to find them and show them the way back. No one need be in any fear of overlooking the ruler of the hive, because she can be known by her "lofty pace and countenance expressing Majesty, and she hath a white spot in her forehead glistering like a Diadem."

An old writer advises that all the hives should have holes bored right through them to prevent spider-webs. He was also of opinion that the bees swarmed because of the queen's tyranny, and if she followed them, they put her to death. He informs us that the drones were honey-bees which had lost their stings and grown fat. This was a very old idea, with which the sceptical Butler dealt in the following fashion: "The general opinion anent the Drone is that he is made of a honey-bee, that hath lost hir sting; which is even as likelie as that a dwarfe, having his guts pulled out, should become a gyant." But the bee-masters of the Middle Ages were ever intolerant of other people's mistaken ideas, while supporting with the gravest argument and show of learning equally benighted superstitions of their own.

A little book published in 1656, and called "The Country Housewife's Garden," is interesting, as it was probably written for cottagers by one almost in the same humble walk of life, whereas the bee-books generally of the sixteenth and seventeenth centuries were, for the most part, the work of men of considerably higher station.

This book, almost alone of its kind, harbours no fine theories on bee-keeping, but keeps throughout to rule-of-thumb methods.

Rev. John Thorley Writing his
"Melissologia" with the Help of his Bees. 1744
(From an old Bee Book)

The writer, evidently caring little for speculation as to the origin of bees, but confining his remarks to practical honey-getting, takes up the following wholesome position: "Much discanting there is of, and about the Master Bees, and of their degrees, order, and Government: but the truth in this point is rather imagined, than demonstrated. There are some conjectures of it, viz., wee see in the combs diverse greater houses than the rest, and we commonly hear the night before they cast, sometimes one Bee, sometimes two or more Bees, give a lowde and severall sound from the rest, and sometimes Bees of greater bodies than the common sort: but what of all this? I leane not on conjectures, but love to set down that I know to be true, and leave these things to them that love to divine." The "greater houses" here mentioned were, no doubt, the large cells in which the queens are bred. Just before swarming-time, as many as nine or ten of these are sometimes to be found in one hive.

The same writer has the inevitable ill word against the drones. These, he says, "are, by all probability and judgement, an idle kind of bees, and wastefull, which have lost their stings, and so being as it were gelded, become idle and great. They hate the bees, and cause them cast the sooner."

Never did creature come by so bad a name, and so undeservedly, as the luckless drone with these old scribes. Another of them speaks of the drone as "a grosse Hive-Bee without sting, which hath beene alwaies reputed a greedy lozell (and therefore hee that is quicke at meat and slow at worke is fitted with this title): for howsoever he brave it with his round velvet cap, his side gowne, his full paunch, and his lowd voice; yet he is but an idle companion, living by the sweat of others' brows. For he worketh not at all, either at home or abroad, and yet spendeth as much as two labourers: you shall never finde his maw without a good drop of the purest nectar. In the heat of the day he flieth abroad, aloft, and about, and that with no small noise, as though he would doe some great act: but it is onely for his pleasure, and to get him a stomach, and then returns he presently to his cheere."

But it is among the writings of the old bee-men with a taste for the quack-doctor's art that some of tne quaintest notions are to be found. We are told that honey, well rubbed into the scalp night and morning, is a sovereign remedy for baldness, and if it was mixed with a few dead bees and a little old comb well pounded, it was still more efficacious. Dead bees, dried and reduced to a powder, form a principal ingredient in all sorts of nostrums of the time. This powder, mixed with water and drunk every morning, is recommended as an unfailing cleanser to the system. And if the heads of a large number of bees are collected, burned, and the ashes compounded with a little honey, it makes an excellent salve for all sorts of eye disorders.

There was a famous preparation called Oxymel, which was in great vogue in mediæval times. It seems to have been nothing more than a mixture of honey, water, and vinegar; but it was accredited with extraordinary virtues. It was an infallible cure for sciatica, gout, and kindred ailments; and one writer also tells us that it was "good to gargarize with in a Squinancy."

But honey and dead bees were not the only products of the hives which were pressed into medical service. Wax also was believed to have exceptional curative powers in all sorts of human ills. It had the faculty of curing ulcers, and "if the quantity of a Pease in Wax be swallowed down of Nurces, it doth dissolve the Milke curdled in the paps." It was also used as an embrocation for stiff joints and aching muscles. The supposed curative value of beeswax in its natural state, however, was as nothing compared to its capabilities when distilled. This preparation, known as Oil of Wax, and famous at the time all the world over, seems to have come nearer the ideal of a panacea—a cure-all—than anything else before or since. The making of Oil of Wax seems to have been a very complicated affair. First the wax had to be melted, poured into sweet wine, and wrung out in the hands. This was done seven times, using fresh wine at each operation. Then the wax was placed in a retort with a quantity of red-brick powder, and carefully distilled. A yellow oil came over into the receiver, and this was distilled a second time, when the "Coelestiall or Divine medicine" was ready. Miraculous portents seem to have accompanied its preparation, for we are told that "in the coming forth of this Oile there appeareth in the Receiver the foure Elements, the Fire, the Aire, the Water, and the Earth, right marvellous to see."

The power to stop immediately the falling out of the hair, heal the most serious wounds in a few days, and cure toothache and pains in the back, can be reckoned only among its minor virtues. Much greater properties were claimed for Oil of Wax, for it not only "killeth worms and cureth palsy and distempered spleens, but it bringeth forth the dead or living child."

One last extract must be given from the same old writer. It relates to the generation of bees, and brings us out, perhaps, on the highest pinnacle of the marvellous. After a learned dissertation on the method of breeding bees from a dead ox—assuring us, however, that if we can procure a dead lion for the purpose, it will be much better, as then the bees will have a lion-like courage—the writer goes on to explain how bees may be produced in another way. We are to save all dead bees, burn them, sprinkle the ashes with wine, and then leave them exposed to the sun in a warm place. In a little while, we are told, all the bees so treated will come to life again, and we shall then have a new stock ready for hiving.

Dipping into these time-worn records of the Middle Ages, with their embrowned, scarce legible type and their antiquated phraseology, one comes at last to realise how very little the old bee-masters actually understood of the true ways of the honey-bee, or, indeed, of any real essential in bee-craft. And yet the production of honey and wax must have been an industry very largely developed in those days. Somehow or other, in spite of archaic theories and useless interference in the work of their hives, these people must have contrived to supply a market of whose magnitude we can nowadays form little conception. The trade in wax alone must have been a very large one, for, except in the poorest tenements, this formed the only available source of artificial light. And honey was in much more universal demand than it is now, because cane-sugar could hardly have developed into a serious rival as a sweetening agent among the masses at a time when it stood, perhaps, at two shillings a pound.

But in speculations of this kind, it must be borne in mind that, although the men who wrote about bees displayed so picturesque an ignorance in all matters appertaining to their charges, these formed a very small minority among the bee-keepers as a whole. Probably the bulk of the supply in honey and wax came from bee-gardens, whose owners neither knew nor cared anything about books, and were concerned only in the practical side of the work, where their knowledge, hereditary for the most part, amply sufficed for the part they played in it.

Moreover, it is only in latter-day, scientific apiculture that the work of the bee-master counts to any great extent. Nowadays, under the light of twentieth-century knowledge, this is competent to bring about the doubling, and even trebling, of the honey-harvest possible under the ancient methods. But the old skeppists did, and could do, little more than look on at the work of their bees, and here and there put a scarce availing hand to it. Nearly all the credit for the results achieved in those days must be given to the bees themselves, who, untold ages before, had brought to finite perfection their remarkable systems and policies. In all likelihood the bee-masters, the practical men who owned the hives, had much the same shrewd faculty of leaving things alone in far-off times as we observe among the skeppists of the last generation. In many ways, what they did at last come to do they did ill, notably in the apparently insane practice of destroying the bees to obtain the honey. But even this was not so foolish a procedure as it appears to-day. It was a plain matter of business, according to the lights of the time. Their process was to condemn to the sulphur-pit all the lightest and the heaviest of their stocks. Experience taught them that the weak colonies stood little chance of getting through the winter unless they were artificially fed; while if the bees of the large colonies were preserved, after being robbed of their stores, they would need the same provision. It was a matter of arithmetic. Artificial feeding was then a much more costly affair than it is to-day, and the reckoning came out well on the side of slaughter. The worst part of the business, so far as modern scientific bee-breeders are concerned, is that the old system of destruction tended to preserve only those strains of bees who were inveterate swarmers; while the steady, industrious stay-at-homes, who accumulated the largest stores of honey, were invariably exterminated. This is a fateful legacy to have passed on, when we consider that one of the chief aims of modern bee-science is to abolish swarming altogether. The swarming habit is one of the greatest obstacles in the way of a large honey yield, and until a race of non-swarming bees has been evolved by modern breeders there will always be this element of uncertainty in the honey harvest.

Latter-day bee-men, therefore, join the chorus of disapproval of this old, senseless custom of bee-burning, rather because it has given them the task of undoing the work of ages before any progress is possible, than from the generally accepted humanitarian reasons.

CHAPTER IV

AT THE CITY GATES

IN a village in Southern Sussex, close under the green brink of the Downs, there live two bee-keepers who represent, in their widely divergent methods and outlook, the extremes of bee-manship as still extant in modern times.

The one dwells in a little ancient thatched cottage, set in the heart of an old-fashioned English garden, where dome-shaped hives of straw are dotted about at random amidst a wild growth of the old-fashioned English flowers. The other has built himself a trim villa on a hillside, topped with a sheltering crest of pine-wood; and here he has established a great modern honey-farm, replete with every device and system of management known to apiarian scientists throughout the two worlds.

One might suppose, on leaving the village street on a fine May morning and coming upon these two settlements in the open country beyond, that all the romance and old-world flavour of bee-keeping were inevitably to be found in the ancient bee-garden, where the droning music of the hives seems to originate in the thicket of blossoming lilac, and red-may, and veronica, the hives themselves being the last things one noticed in such a tangle of bright-hued flowers. To expect sentiment in the other quarter—a great cindered tract of country, with its long parallel rows of modern hives, all painted in various colours, its dwelling-house that might have been transplanted bodily from a well-to-do London suburb, and its line of outbuildings, with their bustle of business, and coughing oil-engine, and reverberation of hammer and saw—was to expect something evidently out-of-date and impossible. As well look for art in a Ghetto as to seek reverence for ancient bee-customs in a twentieth-century trading concern such as this, established to supply the market for honey just as a Manchester factory turns out calico and corduroy.

Many lovers of country life, peripatetic artists and chance pedestrians for the most part, came to the village with this notion firmly impressed upon them, and, visiting the old bee-garden and finding the old beautiful things there in abundance, went no farther, and became no wiser. They wandered round the crooked, red-tiled paths of the garden with its ancient proprietor; stooped under bowers of living gold and purple; waded through seas of scarlet poppy and blue forget-me-not and tawny mignonette; came upon old bee-hives in all sorts of shady, unpremeditated corners; and steeped themselves in mediævalism up to the eyes. The very song of the bees seemed to belong entirely to past days.

None, surely, but a hopeless Vandal could put a colony of bees in one of the ugly square hives, and expect them to go honey-seeking in the old harmonious, happy way, sanctified of the ages. And so they never ventured up the hill to the great bee-farm, but kept to the garden below, and listened by the hour together to the quaint talk of its white-headed, smock-frocked owner, or stood valiantly at the foot of the ladder when he climbed up to dislodge a swarm from the moss-grown apple-boughs, or helped him to scour the new straw skeps with handfuls of mint and lavender, or beat out weird, unskilful music with the door-key on the old brass-pan when a swarm was high in the air.

Much could be learnt, it is true, from quiet days spent in the old bee-garden, especially in May, before the earliest swarms were ready to forsake the hives.

The first faculty to be acquired was that of wandering among the bees, or standing between their straw houses, undismayed at their incessant and often terrifying approaches.

*Inverted Straw Skep-Hive,
Showing Natural Arrangement of Combs*

Whatever confidence one may place in bee-keepers' assertions that their bees never sting, it is a bold man who can preserve entire equanimity when bees are settling continuously on his hands, his face, his clothing, and a whole flying squadron of them are shrilling vindictively about his ears. Nothing will come of it, he knows, if only he can keep still. But the tendency to turn and flee, or at least to beat off these minatory atoms with wildly waving arms, is all but irresistible for the novice. It is only their way, he is assured, of expressing or of satisfying their curiosity; and, this being done, they fly off harmlessly enough to give a good report of him to the ruling powers within the hive. But he knows that this report is sometimes anything but good. At least, there are a few luckless individuals in the world who dare not venture within a dozen yards of a bee-hive without being set upon unmercifully, and chased by an angry squad of these tart virgins for the space of a quarter-mile. Moreover, in certain states of the weather—when thunder is about, and the air is tense and still—bees will often sheath their barbed daggers in any human skin, even that of their owner, who has gone among them daily all the season unmolested. There is, therefore, a fateful element of chance in all near watching of bee-hives, a sensation of being under fire—fine discipline enough, but, for the timorous, hardly to be reckoned among the easy joys of existence.

These first deterrents, however, being happily overcome, the watcher is sure to be caught up, sooner or later, in the sheer fascination of the thing, and to find himself recklessly, almost breathlessly, looking on at what is nothing else than a great informing pageant of life.

He stands, as it were, a stranger at the gates of a city, inhabited by the most interesting, and in some respects the most advanced, people in the world. Of the inner life of the city, apart from the deep busy murmur that surges out to him, he learns nothing, and will learn nothing until he puts sentimental pride in his pocket, and makes pilgrimage to the great bee-farm on the hill. But here, in the meanwhile, is food enough to satisfy the keenest appetite for the marvellous. In and out through the yawning entrance-gate of the city, under the hot May sunshine, there are thousands of busy people coming and going. The broad threshold of the hive is completely hidden under opposing streams, the one setting out towards the fragrant fields and hedgerows, the other tumbling and seething in, almost every bee dragging after her some kind of mysterious treasure.

The outgoing bees start on their journey in two different fashions. Some emerge from the hive and rise at once on the wing, lancing straight off into the sunshine; and these are foragers, who have already made several journeys afield since the sun broke, hot and rosy, over the eastward hill. But others, essaying their first excursion for the day, creep out of the murmurous darkness of the hive, and come with a little impetuous rush to the edge of the alighting-board. Here they pause a moment to flutter their wings and rub their great eyes free of the hive-twilight. And then they lift into the air, hover an instant with their heads towards their dwelling, taking careful stock of it, sweep up into the blue, and volley away with the rest towards the distant hill-side, white with its bridal wreath of clover-bloom.

The homing bees move much more sedately. They come sailing in like bronze argosies laden to the water's edge. Those bearing full sacs of clover-juice for the honey-making seldom carry an outside load of pollen as well. They have all to do in bringing their distended bodies to a safe anchorage on the entrance-board, and charge headlong into the hive, possessed of only one idea— to hand their garnered sweets over to the first house-bee they chance upon, and then to hurry out in search of another load. The pollen-bearers are impelled by the same white-hot energy; but their cargoes are infinitely more cumbersome, and demand a more leisurely pace. Some with panniers, heaped up with a deep orange-coloured material, must rest awhile on the threshold before gathering energy enough to drag their glowing burdens through the city gate. Others just fail to make the harbour, and sink down on to the grass below, to wait for the same freshet of strength that is finally to bring them into the security of the populous haven. Scores of them do not try for harbour at first tack, but, coming safely into the calm waters of the garden, rest awhile on the nearest leaf or blossom, panting and tremulous, until they are able to wear sail for the last reach home.

There is infinite diversity in the loads of these pollen-carrying bees. Hardly a colour, or shade of colour, in the rainbow fails to pass during every moment across the thronging way. Every bee carries a half-globe of this substance, beautifully rounded and shaped, on each of her two hind-legs. It is possible, by marking the colour of her burden, to tell with certainty what flower she has been plundering on each of her trips. This bright orange, which makes always the largest and heaviest bales in the stream of merchandise, is from the dandelions. From the gorse-flowers come loads of deep rich brown almost as weighty. The charlock, that mingles its useless, wanton beauty with every farm-crop, yields the bee interminable gold. White clover, red clover, sainfoin, all load up the little hive coolies with different shades of russet. From the apple-orchards come bursting panniers of pale yellow; the blackberry-blossom yields pollen of a delicate greenish-white. When summer comes, and the poppies make scarlet undertones amidst the wheat and barley, these winged merchant-women stream homeward with their pollen-baskets laden with funereal black.

But, if you watch a hive at work on any bright spring or summer morning, you will see single bees occasionally pass with loads whose source has never yet been fathomed. The lean, glistening, rufous stuff that is continually borne through the hustling crowd is resin gathered from poplar or pine, and used to glue the straw hive down to its base-board, or to stop up draughty crevices and useless corners, or, diluted into varnish, to paint the honeycombs with an acid-proof, preservative film.

But now and then comes a bee with a load whose colour shines up like a danger-signal in darkness. Brilliant scarlet, or soft rose-crimson, or pale lavender, or gleaming white—who shall say in what far, forgotten nook of the country-side she has been adventuring, or what rare blossom she has chanced upon in the wilderness, and, despoiling it of its maiden treasure greedily, has quickened into duplication the beauty that was its reason for life?

An Old Sussex Bee-House

Yet the greatest wonder about all this pollen-gathering is that each separate load has been taken entirely from one species of flower. The little half-spheres are packed into the pollen-cells indiscriminately, orange on brown, pale yellow mingled with green, or buff, or grey. But each pair of panniers, representing a single journey, contains the pollen-dust of one kind of blossom alone. Going out into an English lane or meadow to watch the bees at work, the first conviction borne in upon an observer is that the bees are darting about from flower to flower without other thought than to load up from any and every capable blossom that stands in their way. But closer scrutiny reveals a curious plan and order in this, as in everything else that the honey-bee undertakes. Tracing an individual bee in her progress along the flowery verge of the lane, you will soon see that she visits only one species of blossom. If she starts on hawthorn, it will be hawthorn all the way. If her load of willowherb-nectar or pollen is not yet a full one, she will overpass a score of tansy-knots or waving jungles of meadow-sweet, just as inviting and resourceful, apparently, to reach the one scanty patch of purple at the end of the lane.

Why she should be at such pains to keep the pollen separate as she gathers it, only to get it inextricably mingled with every other kind in the storehouse at home, is a problem that none but a bee can solve. But all the honey-bee's reasons and motives in life are made up of a curious blend of cold-drawn sense and sentiment; and it may be inferred that need and fancy have an equal influence in guiding her in this, as in everything else she does, from her cradle-cell to her grave. Not altogether without seriousness, it may be hazarded that quite as probable a reason for her way of pollen-gathering is that she deems a certain shade of colour makes a more becoming flying-robe, as that she keeps each load of pollen pure, unblended, because of some imperious, economic need of the hive. The factor of sex, in all observation of the ways of the honey-bee, is no more to be considered a negligible one than it is in the critical contemplation of the human species of hive.

All this incessant coming and going of the busy foragers is alluring enough to the looker-on, but there is evidence of many other activities equally interesting. The work of collecting nectar and pollen is obviously only a part of the duties of this self-immolated spinster-race. Here and there in the seething, hurrying crowd there are bees wha do not move with the rest, but, anchored securely in the full force of the living current, with heads lowered and turned towards the hive, are engaged in fanning their wings, and this so swiftly that nothing of the wing but a little grey mist can be seen. Looking more carefully, you will make out that these bees are arranged in nearly regular rows, one behind the other, in open order, so that the conflicting tides of foragers can pass uninterruptedly between. If the watcher is bold enough to bring his ear down to the level of the hive, he will make out a steady hissing noise that rings clear above all the din and turmoil made by the incessant travellers to and fro. These rows of fanners are seen to stretch from the hive-door right to the edge of the footplate, but principally on one side; and still closer observation will reveal the fact that there is a regular system of relief among them. Though the general volume of sound never abates one jot, every few minutes one or another of these stationary bees moves away, her place being immediately taken by another, who settles down to the common task in line with the rest. The reason for all this is plain enough: the fanners are engaged in ventilating the hive, drawing a current of vitiated air through the entrance on one side, which flanks, but does not oppose, a corresponding current of pure air sucked in on the other.

All through the warm days of spring and summer this fanning squadron is constantly at work; nor does it cease with the darkness. Chill nights find the ranks weakened and reduced to perhaps only a few bees, or even to none at all when a cold snap of weather intervenes. But in the dog-days, or, as the ancients used to say, when Sirius, the honey-star, is shining, the deep sibilant note of these fanners rises, in a populous apiary, almost to the voice-strength of a gale of wind. To come out then under the stars of a summer night, and stand listening in the tense, fragrant darkness to this mighty note, is to get an impression of bee-life unattainable at any other season. In the daytime the sound is intermingled, overwhelmed, by the chorus of the flying bees. But now all are safely at home. Each hive is packed from floor to roof with tens of thousands of breathing, heat-producing creatures: the necessity for ventilation is quadrupled, and, far and wide in the bee-garden, the fanning armies are setting to their work with a will.

The freshman at this fascinating branch of nature-study, brought out into the quiet night to hear such gargantuan music, is always strangely affected by it, some natures incredibly so. In all the great placid void of darkened hill and dale around him, in the whole blue arch overhead, alive with the flinching silver of the stars, there is no sound but a chance trill of a nightingale, the bark of a shepherd's dog on the distant upland, or, now and then, the droning song of a beetle passing invisibly by. All the world seems at rest, save these mysterious people in the hives; and with them the sound of labour is only redoubled. Bending down to the nearest hive in the darkness, the note comes up to one like the angry roar of the sea. A light brought cautiously to bear upon it, discloses the alighting-board covered with rows of bees, working, as it were, for their lives; while other bees continually wander in and out of the entrance—the sentries that guard it night and day, just as soldiers guarded the gates of human cities in olden times. The novice at bee-craft, even the most staid and matter-of-fact, is invariably plunged into marvelling silence at the sight. But if the night be exceptionally hot and oppressive, and the fanning army unusually large, the bee-master with an eye for dramatic effect generally finishes the tiro's wonderment by showing him an old trick. He lowers the candle until the flame is just behind the squadron of ventilating bees, and at once all is darkness: the current of air drawn out of the hive has proved strong enough to extinguish the light.

It has been said that there are guard-bees who watch the hive-door day and night. To the unskilled human eye one bee looks very like another, and it is difficult to understand how, in the many thousands that pass, the guards manage to detect an intruder so unerringly, and to eject her with such unceremonious promptitude as is always shown. Probably it is not by sight alone that these occasional interlopers are singled out. The sense of smell in the honey-bee is extraordinarily acute, and this, no doubt, assists the guards in their difficult work. It is well known that a queen-bee must possess a very distinct odour, as her mere presence abroad, even when shut up in a box, will attract the drones from all quarters. In all likelihood the peculiar aroma from each queen-bee impregnates the whole colony, and thus the guard-bees are able at once to distinguish their own kin from that of alien stocks.

Still watching the outside life of the hive in the old bee-garden, many other interesting things come to light. In such an establishment, even if it be only an old-fashioned straw skep, perhaps more than twenty thousand individuals are located; and obviously some regular system of cleaning and scavenging is indispensable. This work can be seen now, going on uninterruptedly in the midst of all the other busy enterprises. Every moment bees come labouring out, bearing particles of refuse, which they throw over the edge of the foot-board, and at once shoulder their way back for another load. Other bees appear, carrying the bodies of comrades who have died in the hive; and every now and then one comes struggling through the crowd, bearing high above her a strange and ghastly thing, perfect replica of herself, but white throughout, save for its black beady eyes. This is the unborn bee, dead in its cradle-cell. Infant mortality is an evil not yet overcome even by the doughty honey-bee, and many are carried out thus, especially in early spring. Watching these undertakers of the hive in their gruesome but necessary work, a singular fact can be noted. While all other debris is merely cast over the brink of the entrance-board, where it accumulates day by day on the grass below, these dead larvæ are never disposed of thus. They are carried right away, their bearers taking wing and flying straight off over the hedgerow, to drop them at harmless distance from the neighbourhood of the hive.

There is still another kind of work going briskly forward round the gates of the bee-city. Certain among these stay-at-home bees seem to exercise a sort of common overseership. They help those weighed down with too heavy a cargo to reach the city gates. If a lump of pollen is dropped in the general scuffle, these bees seize it and take it into the hive. Sometimes a bee comes eddying downward, smothered from head to foot with pollen, like a golden miller, and she is immediately pounced upon by these superintendents, and combed free of her incommodious treasure. Others see to the grooming of the young bees, about to essay their first flight. The youngster sits up, protruding her tongue to its fullest extent, while half a dozen bees gather round her, licking and stroking her on every side. At last her toilette is done, and she is liberated, when, with a little flutter of her wings, she lifts high into the blue air and sunshine and makes off with the rest to the clover-fields, glittering afar off in the joyous midday light.

For insensibly the hours have worn on—it is noon—and the tense thronging life, the deep rich labour-song, of the bee-garden seem to have reached their height. But suddenly a greater noise than ever arises on all sides: a steady stream of bees, larger and bulkier than the rest, is pouring out of every hive. The drones, the lazy brothers of these laborious vestals, have roused at last from their sleep, and are coming abroad for their daily flight. In twos and threes, in whole battalions, they hustle out, and begin their noontide gambols about the hive, filling the air with a gay, roistering song. In a little while they will be all gone to their revels, and the bee-garden will seem, by comparison, strangely quiet. But now the sudden accession of energy is unmistakable. With the awakening of the drones there seems to be a new spirit abroad. The air is no longer filled to overflowing with busy foragers. Many of these have joined the dance round the hives, so that each bee-dwelling is the centre of a singing, gambolling crowd, moved rather by a spirit of play, almost of idleness. But this brief moment of relaxation soon passes. The drones betake themselves to their marital pleasuring in the fields. The noisy midday symphony dies down to the old steady monotone of work. And the watcher at the gates of the bee-city turns to retrace his steps down the flower-garlanded way of the old pleasance, satiated with wonders, yet not satisfied, his curiosity only quickened a thousandfold for that which has been inexorably held from him, a glimpse of what is happening behind those baffling walls of straw.

Wending slowly homeward, and pondering, he asks himself many questions. What is the reason, the final outcome, of all this earnest, well-directed labour? What is done with the pollen that has been carried in all the morning long? Where there is obviously so much system, and unanimity, and ingenious division of endeavour, there cannot fail to be a supreme and governing intelligence to allot the part that each must play. This story of a queen—of a single bee, larger than all the rest, to whom all pay allegiance, and who spends her whole life in the dim labyrinth of the hive, like the Pope in the Vatican—is it a truth, or only a figment of the ignorant, bucolic brain? If this queen exist, if every hive have indeed its absolute monarch, who directs the whole complex life and policy of the bee-city, where in the scale of reasoning creatures must she be placed?

The City Gate

And then, if he be wise, the student will learn at last to give the picturesque old bee-garden its true appraisement. Ancient things conserve their beauty, and win the love of the right kind of lovers, more and more with every century that glides by. Only their usefulness, their import in the tide of human knowledge and progress, has gone with the years. It is so with the bee-garden under its Maytide robe of green leaves and rainbow blossoms.

It is beautiful in its glad appearances, its echo of old voices, its odour of the sanctity in ancient ways and days. But it can tell us nothing of all we want to know. It can only ask us riddles to which we have no answers. For these we must set aside old fanciful scruples; turn our backs, once for all, on its enchantment and its sweetness; bend our steps unswervingly towards the great modern bee-farm on the hill.

CHAPTER V

THE COMMONWEALTH OF THE HIVE

A DOCTOR DRYASDUST will manage tc impart to the truths he meddles with a disastrous air of dulness and stagnation; but to walk in a fools' paradise of beautiful, artistic error is to lay oneself open to an infinitely worse fate. There never was a truth in Nature that was dull or uninteresting, except in its human presentment. There never was a pretty worthless fiction that did not show its dross and tinsel when brought out into the searching light of day. Romance, the spirit of poetry, have largely changed their venue of recent years. The unconscionable delver among old things, old thoughts, old conventions, on the strand of Time, has tarried so long in his one little florid corner that he is in some danger of being caught by the tide. He must soon either mend his pace or swim for it. Human regard is turning more and more towards those who deal in living verities—the men who search the stars, who win new powers out of the common air, who find at last the authentic teachings in the old worn texts of the stones and brooks. These are the true poets, romancists, tellers of wondrous tales; and these will hold the crowd—which is never far astray in its intuitions—when all the singers of sick fancies and the harpers on frayed golden strings have gone off in melancholy dudgeon to their own place.

The old story—which has held such a long and honoured position in school text-books, and in the writings of those who tell of Nature's wonders from the commanding watch-tower of the study fire—the old story of the queen-bee ruling her thirty or forty thousand dutiful subjects, and guiding them unerringly in all their marvellous exploits and enterprises, must go now with the rest. For the truth, as modern observers have unquestionably established it, is that the queen-bee is no ruler in the hive, but even a more obedient subject than any. The real instigators and contrivers of everything that takes place within the hive are the worker-bees themselves. The queen has neither part nor lot in the direction of the common polity; nor has she any power, mental or physical, to help in the carrying out of public works. Her sole duty is that of motherhood, and even in this she derives all initiative from the sovereign worker-bees. She is little more than an ingenious piece of mechanism, and carefully guarded and cherished accordingly. She has certain propensities, and certain elemental passions, which she can always be counted on to exercise in certain well-defined and limited ways. But as an intelligent, originating force she counts for nothing. The mind in the hive is the collective mind of the whole colony, apart from the queen and drones—an hereditary, communal intellect evolved through the ages, the sum and total of all bee experience since the world of bees began.

If, however, modern science compels us to divest the mother-bee of all her regal state and quality, and thus destroy one of the prettiest delusions of ancient times, it is only to take up a story of real life more alluring and romantic still. In the light of new understanding the old facts take on a mystery and excite a wonderment greater than ever before. If we found the life of the hive an enthralling study when we supposed it to originate from one winged atom endowed with acute and commanding abilities, how much more fascinating must it prove when we come to see that all this complex system of government is instituted and kept together by the harmonious working of tens of thousands of reasoning beings?

Reasoning—it is a big word, a double-edged thing that requires careful handling. We have been so long accustomed to use it only in regard to our own magnificent mental processes that it savours almost of the ridiculous to bring it to bear upon such a tiny et-cetera in the brute creation as the honey-bee. And yet, the deeper we go in the study of the bee and all her works, the more difficult it becomes to find a word that shall more fittingly meet the case. Instinct will not do. Instinct implies a dead perfection of motive, born of omniscience, working through unthinking, unvarying organisms to an equally perfect end. But in neither project nor performance can the honey-bee be said invariably to achieve, or even to aim at, perfection. It will be seen hereafter that her motives, her methods, the results she brings about, all show frequent, undeniable error or deviation. She attempts to carry through a sound enterprise, but abandons it on finding unforeseen difficulties in the way. She will persevere blindly in an obviously foolish piece of business, and fail to see her mistake until both energy and resources are at an end. Sudden emergencies may find her ready with the saving stroke of last ingenuity, or merely plunge her into listless despair. Courage, industry, economy, wise forethought, or still wiser afterthought, are all common traits in her nature. But she may develop idleness, unthrift, slovenliness, or even downright dishonesty, if chance or circumstance indicate the way.

And what are all these but the defects or attributes of reason? If bees and men, each admittedly rooted in divinity, be prone to the like failings and inconsequences, who shall discriminate between them, dividing arbitrarily natural cause and effect?

Watching bees at work for the first time through the glass panels of an observation hive, or in the almost equally informing modern hive with movable combs, this question continually arises, and there seems only one answer for it. There is something curiously human-like in their movements over the crowded combs, and the old comparison of a bee-hive to a city of men is never out of mind. There are the incessant hurryings to and fro; chance meetings of friends at odd street-corners; altercations where we can almost hear the surly complaint and tart reply; busy masons and tilers and warehouse-hands at work everywhere: a hundred different enterprises going forward in every thronging thoroughfare or narrow side-lane, from the great main entrance to the remotest drone-haunted corner of the hive.

You will see the huge, full-bodied queen labouring over the combs from cell to cell, with a circle of attendants ever about her. In the highest stories of the hive the honey-makers are at work, pouring the new-garnered sweets into the vats, or sealing over with impervious wax the mature honey. Where the nurseries are established, in the central and warmest region of the hive, the nurse-bees are hurrying incessantly over the combs, looking into each cell to mark the progress of the larvæ; giving each its due ration of bee-milk; or, when the time arrives, walling up the cell with a covering that shall insure its privacy, but freely admit the air. Here and there the young bees have awakened from their transforming slumber, and are clamouring at the stoppings of their prenatal tombs, gnawing their way out vigorously, or thrusting forth red, glistening, ravenous tongues, eager to end their long fast. Where these raw youngsters have at last won their way into existence, they can be seen assiduously grooming themselves, or searching the neighbouring comb for honey, while the nurse-bees are busy cleaning out the cells, just vacated, to make them ready for the queen when she comes by on her next egg-laying round.

And all these operations are going forward simultaneously on an incredibly large scale. Certain amazing scraps of information are given to the wondering on-looker, which he hears, but can, at this stage in his progress, seldom rightly estimate. He is told that the queen is the only mother-bee in the colony, large as it is; that, in the prime of her maternity, she will lay as many as 3,000 eggs a day; and that she has the power to produce either male or female eggs, or none at all, at will. He is told that, except when she leads forth the swarm, she goes out of the hive only once in her life, and this is her wedding-trip. On this one occasion she has traffic with the drone somewhere incredibly high up in the blue air and sunshine of the summer's day; and that immediate death is her suitor's invariable portion; that she returns at once to the hive, and thereafter for the rest of her life, which may endure for years, she passes her time in immaculate widowhood, yet retaining her fertility to the end.

Comb-Frame from Modern Hive, with Queen

She is pointed out to the gaping novice as she travels her unceasing round of the brood-combs, and her various attributes are explained to him. He is shown how much larger she is than the worker-bee; how her bodily structure differs in a dozen important ways; how her instincts and habits resemble those of the common worker hardly in a single particular. Finally he is told something at which the most polite credulity may well demur. Although the mother-bee is to all appearances of a totally different race, the egg from which she was raised was identical with that which produces the little worker. Her bodily size, the change in the number and shape of her organs, her mental differences, are all due to treatment and diet alone. There is no reason why she should not have been an ordinary neuter working-bee, nor why any one of the thirty or forty thousand little workers in a hive should not have become a great queen-bee, the sole mother of an entire colony, save for the edict of the communal mind. More wonderful still, the drones, the male bees—the brothers, never the fathers, of their own hive, as has been so often stated—owe the fact of their sex entirely to the will or whim of the hive authorities, working through the docile agency of the queen. Until the moment before the egg is laid, the question of the sex of the resulting bee is held in abeyance. This big lusty drone, with exuberant masculinity obvious in every posture and act; his totally different organism; his incapacity for anything else than the fulfilment of the one office required of him, for he cannot even entirely feed himself; his habit of spending his life either in a comfortable lethargy of repletion at home, or in amorous knight-errantry abroad—this drone might have been a little plodding worker-bee, with shrunken yet elaborated body and curiously developed brain, whose one idea in life is to get through the largest amount of work before death claims her, and who is armed with a formidable poisoned sting, while the drone has none.

It is useless at this stage to tell the learner that all these vital differences—miracles, indeed, in the ordinary meaning of the word—are brought about by the leading powers of the hive in certain simple, easily explainable ways. He has lost, for the moment, all sight of and interest in the details, however extraordinary, in the perception that has dawned on him of the vastness of the entire plan. Here is a community that, to all appearances, has solved every problem relating to the well-being and progress of a crowded, highly organised society. Questions that are now vexing socialistic philosophers in the human world, or are looming dark in the immediate future—problems of numerical increase in relation to food-supply, the balance of the sexes, communal or individual ownership in property, due qualification for parenthood, the hegemony of might or right—all seem to have been happily settled long ago in this remarkable bee-commonwealth. In itself a prosperous, well-conducted hive appears to offer a living example, a perfect object-lesson of what Socialism, carried out to its last and sternest conclusions, must mean to human and apiarian communities alike. Here is a number of individuals—counting anything from ten thousand to fifty or sixty thousand, according to their condition and the time of year—living heathily and comfortably in the space of a few cubic feet. The principle, all for the greatest good of the greatest number, is elevated into a prime maxim, to which every one must bow. The fiction of royalty is maintained in harmony with the perfect republican spirit. The females are supreme in everything, the males in nothing. Growth of population is accelerated or retarded, according to estimations of the immediate or future supply of food. The proportion of the sexes is varied at will. The rule, that those who cannot work must not live, is applied with relentless consistency. All the garnered wealth of the State is held in common for the common good. When the settlement becomes too populous, and the boundaries cannot be extended, a large part of its inhabitants are forced to emigrate, taking with them only so much of the state property as they can carry in their haversacks, and relinquishing all claim to the rest. The governing females have apparently agreed among themselves that only one of their number shall exercise the privilege of motherhood; and when her fertility declines, she is deposed, and a new mother-bee, specially raised for the purpose, installed in her place.

All these, and a host of other facts as to bee-life, are crowded into the bewildered brain of the tiro until its capacity is exhausted, and he can take no more. He begins to see, at length, that he is approaching a great matter too fast, and from the wrong direction. Like a scholar who, resolving on a new and difficult branch of study, commences at the end of his treatise instead of at the beginning, he finds himself in the midst of terms and equations of which he knows nothing. All this desultory peering into hive-windows, and listening to scraps of astounding information, is nothing but opening the book of bee-life here and there at odd disjointed pages, getting a swift impression of certain lurid, kaleidoscopic details, but no grounding in the consecutive science of the facts. There is nothing for it—if he be resolved to know the life of the honey-bee truly—but to turn back to the first page of the volume, and steadily work his way through to the end—if end there be.

All know the English honey-bee—the Black Bee, as she is called, partly to distinguish her from her foreign rivals, and partly, it would seem, because she is not black at all, but a rich brown—but all do not know her origin. Probably she came to us from the tropics by easy stages, swarm out-flying swarm, until the most adventurous crossed the English Channel in remote ages, when it was only a narrow race of water, or even before Great Britain was detached from the mainland.

It was the black bee, and not the motley-coloured Italian or other varieties, who came to us thus, for the same reason, probably, that the Celts came—because they were a hardy race, loving, and being more fitted for the bracing northern atmosphere than the heat and languor of the south. Modern bee-breeders who are trying so hard to acclimatise in Britain the golden-girdled or silver-fringed bee-races of other lands, might well ponder this fact. No keener controversy rages to-day among English bee-masters than this one of the relative merits of native and foreign stocks. But assuredly Nature has not erred in this respect. South Down sheep can be reared in any county, but nowhere so fine as on the Sussex Downs. The like principle holds good with the English bee. The ages have evolved her from her tropic beginnings to make her what she is—a doughty, essentially British creature, thriving against all odds of fickle climate, when her more tender sisters from the south are hard put to it for a living. She has held her own against them, and more than her own. In bumper seasons, such as we get all too rarely, when, in sober truth, the land is flowing with honey, there is little to choose between the rival honey-makers. But through good and bad, early and late, for steady, dogged industry, invincible hardihood, tangible results, the English black bee has out-distanced all competitors. Thousands of years have gone to her making, and thousands more may conceivably fit the yellow-skirted Ligurian for British work. But labour for so remote a posterity were altruism meeter for angels than for men.

In her old primæval fastnesses the honey-bee is little likely to have troubled herself with hive-making, but to have hung her combs to some convenient branch in the forest, much as the bees in India do to-day. The habit of seeking some hollow tree or cleft in the rock grew upon her probably as she advanced northward, and some nightly or seasonal shelter became more and more an imperious need. The present-day customs of wild creatures give some inkling of their ancestral ways, but it is in their occasional aberrations from these customs that we get the truest indications of what their original state must have been. Lost swarms of bees, if they fail to pitch upon some better site, will often build in the open, either suspending their waxen houses from some horizontal branch, or making them in the heart of a thick bush.

The ways of the honey-bee are full of such deviations, due, perhaps, to the working of old ancestral memory rousing dimly in the midst of modern needs. The issue of a swarm may be nothing else than the survival of an old process, vital enough in its day, but, under the present civilised conditions of bee-life, lacking the whet of entire necessity. For, in all respects, the life of the bee, ancient as it is, is an evolved civilisation, and not a surviving, aboriginal state. It is conceivable that the foxes have their holes, and the birds their nests, much after the same fashion as in the days when Adam invented love-making. But the twentieth-century honey-bee is not of this kind. The communal habit itself may even have been a comparatively late introduction in her progress. It is possible to get some idea of the path she won for herself through the ages by studying the ways of creatures now living, but immeasurably less advanced than the bee. There are distant connections of hers—lonely little wood-wasps and others—which never associate with their kind, but get through the short summer hours in solitude, and die with the waning season, leaving the perpetuation of their species to the children they never see. The common wasp is nearer the honey-bee in development, but still infinitely far behind. The fecundated queen-wasp comes out of her winter hiding-place, fashions a cell or two in some hole in the ground, and deposits her first eggs, thus laying the foundation of a colony which, populous enough in the season, must nevertheless perish with the next winter chills.

In the primæval tropics the honey-bees may have lived in separate families, each with its teeming mother, its indolent, lie-abed father—the Turveydrop of creation—and its bevy of youngsters, every one going out, when grown, to establish a home for itself. The modern bee-city, with its complicated systems and laws, and its innumerable multitudes, may have originated only when change of habitat and climate brought about the necessity for a new order of things. Living in perpetual warmth, in a land where blossom followed blossom in unending succession, there would be no need for such co-operation. The one little family, snugging close in its moss-roofed corner, could sustain its own temperature; and where there was unceasing array of nectar-producing flowers, foresight would have been folly: the winter larder would have been left to take care of itself.

But as the young bees, leaving their homes, and flying ever northward, came first into temperate zones, and then into the fringe of Arctic influences, the conditions gradually changed. The perpetual sipping-garden was left behind; and a season came in each year—short at first, but inevitably lengthening—when there were no flowers. Hard necessity must have taught the bee, then, first to gather together with her kind for warmth during the cold season; and then, as this got longer and longer, to make some food-provision for winter days that would eke out endurance until the spring sun again wooed the earth into flower-giving. Thus the first communal bee-nests must have been evolved from the universal need of the race: the first common storehouses instituted: a host of unforeseen difficulties and side-issues encountered, and means for dealing with them contrived. The spirit of invention must have been busy then with the race, and taxed to the limit, of her resources. For never did Pandora open celestial casket upon earth with more redoubtable consequences, than when the Great Artificer set up the honey-bee as an examplar of city-building to the nomadic world of men.

From the crowding together of the separate bee-families for mutual protection against the elements, to a complete and permanent fusion of life and interests, must have been only a step, as Nature works. But then there must have been stirring times—social upheavals, educative disasters, a cataclysmic war of sex. Bee-life must have been shaken to its very foundations. When and how the woman-bee first got the upper hand in the direction of affairs, it is unimportant to determine. But it is certain that she got it, and has kept it ever since. The population problem must have been the great, overwhelming one. With hundreds of prolific mothers in the hive, each having enough to do at home in rearing her own children, and a crowd of lazy, irresponsible drones who could do nothing but dance in the sunshine or go a-wooing, how were the daily needs of the hive to be satisfied, leaving out of account the provision that must be made for coming winter days? It was clearly a case of reform or annihilation; and it may be conceived that the woman-bees, in default of masculine initiative, took the reins into their own hands.

It is a prophetic story. First they discovered their latent powers. The harmless ovipositor revealed itself as a prime weapon of offence. Thus the army was with the revolutionaries, and the rest was easy. A great, far-reaching scheme was set afoot. Motherhood was to be a privilege of the few and the fittest; work the compulsory lot of the mass. Hard times had already bred a lean, unfertile gang among them, and it was discovered that famine rations in the nursery meant a wholesale increase in these natural spinsters of the race. Henceforth the little sex-atrophied worker-bee was multiplied in the hive, while the fully nurtured mothers were gradually reduced to a few—at last to one alone. It was a triumph of collective self-sacrifice for the well-being and high persistence of the race.

All this may be imagined as having taken place in infinitely remote times, long before man succeeded in distinguishing himself from the apes. In the honey-bee of to-day, and her life in the modern hive, we get a sort of quintessence of the ages; a creature developed in mind and body by her unique conditions, these conditions again imposing upon her unique systems of life. Like Ruskin's Venetian, she must live nobly or perish. Much more is required of her than the rôle of domestic and political economist. To make the modern bee-hive a possibility there must be architects, mathematicians, and chemists within its walls. Sanitary science must have its skilled exponents, or the hive would change into a deathtrap within a few hours. There must be land-surveyors ready to explore the country, just before the issue of the swarms, to determine for them their new location. There must be overseers, gang-forewomen, everywhere to superintend every work in progress throughout the hive. Above all, there must be a supreme central power, a far-seeing intelligence, to divine the imminent common need, and to set the forces of the State to work, in right time and order, to provide for it. If all these cannot be proved to exist in a hive of bees to-day, at least the necessity for them is undeniable; and as undeniable, the achieved results.

CHAPTER VI

EARLY WORK IN THE BEE-CITY

THE "turn of the days," when the winter sun has passed its nadir of feebleness and just made its earliest wan recovery in the skies, marks the true beginning of the honey-bee's year. Then the first few eggs are laid in the heart of the brood-nest; the drowsy cluster begins to show an interest in life; the water-carriers bestir themselves, watching for a bright warm morning that they may sally forth to ply their trade.

Dangerous work it is at this season, yet most necessary. Without water the rearing of the young bees is impossible on any but the smallest scale. Water is needed at every stage of their development, and, lacking it, the progress of the colony must be fatally checked. Even the mature bees will starve and die in the midst of plenty, if their honey-stores are candied, and no water is available to dissolve the inassimilable sweets. The hive that shows honey crystals thrown down on the floor, and littering the entrance, is sure to be in desperate case. The bees are tearing open every store-cell, casting away the solidified honey as refuse, to get at the moister portion below. If the cold spell does not break, or the bee-master is unready with his artificial supplies, the colony must perish. So the water-bearers watch for the sunshine, and its first warm glance brings them out to rifle the nearest dewdrops, or track down by its bubbling music the hidden woodland stream. Many die at this work in the early months of the year, chilled by their load on the homeward journey, or snapped up by hungry birds. But at every cost the future life of the colony must be assured, though, of all the hive-people, none but the queen-mother will be alive to see it in its summer fulness.

We are accustomed to think of a hive of bees as a permanent institution, Death playing his old, unceasing, busy part, but young Life more than outplaying him, just as the way is in a city-hive of men. The analogy holds good, but in bee-life the changes are infinitely more rapid. The life of the worker-bee extends, at most, to six months or so; and in the busy season she may die, worn out by labour, in as many weeks. The reapers of last year's honey-harvest were dead by the autumn. The late-born bees, that went into winter quarters with polished thorax and ragged wings, survived only long enough to nurture their immediate successors; and these, again, will live but to bring to maturity the young spring-broods. Not a bee among them will ever again go honey-gathering. Except for the long-lived queen-mother, and the old hive and its furniture, each colony with every year becomes a totally new thing.

Hibernation, in the true sense of the word, has no part in bee-life. The queen-wasp and countless other creatures hibernate, passing the cold months in a torpor of sleep until the enduring warmth of another year lures them back to active existence. But the honey-bees have a better way: they gather together in a dense, all but motionless cluster in the heart of the hive, with their precious queen in their midst and their food-stores above them. At this time honey is their only necessary food, and very little of this suffices to keep up the needful temperature of the colony. When they are out and about at their work, or busy within the hive, the nitrogenous pollen must be added to their daily ration of nectar to build up wasted tissues; but now honey, the nectar concentrated, the heat-producer, is all they want. The bees of the cluster nearest to the combs broach the full cells beneath them, and the honey is passed through the crowd, each bee getting its scanty dole.

Economy is now reduced to a fine art. None knows when a fresh supply may be available, although no chance will be lost to replenish the larder at the first sign of returning warmth. But now the barest minimum of food is taken, and as the nearest cells become emptied of their contents, the cluster moves a step upward. Thus there is a system of slow browsing over the combs, until the dense flock of bees has reached the highest limit of the hive, when new grazing-ground must be taken. But the movement of the cluster is exceedingly slow, perhaps the slowest thing in the animate world. All recognise that existence depends on the stores being eked out to their uttermost. It is a scientific damping-down of the fires of life—a carefully thought-out and perfected plan for preserving the greatest possible number of worker-bees alive on the smallest practicable amount of food, so that the largest possible army of nurse-bees and foragers may be at hand in the springtime to raise the young bees that are to represent the future colony.

But there is no hibernation. It is doubtful even if bees ever sleep, either in their season of greatest activity or in the coldest depths of winter. At all times a slight rap on the hive will awaken an immediate timorous outcry within. Sturdy knocking will soon bring the guard-bees to the entrance to find out the cause of the disturbance, and many bees lose their lives from this vigilant habit alone. On frosty days the tits may often be seen perched on the entrance-board of a hive, beating out a noisy tattoo, and snapping up every bee that emerges; and many other small birds have discovered the same never-failing source of a meal.

The fact that, with a healthy stock of bees, the interior of a hive always preserves its clean condition, is usually a great puzzle to the novice. In the summer, when the bees are passing continually in and out, this is not so vast a matter for wonder. But in winter-time, when the colony is confined to the hive often for weeks together, it is remarkable that neither the combs nor floor of the hive are ever soiled by excreta. This is a difficulty that the sanitary department in the hive has successfully coped with long ago. It must have been one of the earliest problems that presented itself when the honey-bee first evolved the communal habit. The Ancients believed that all the excreta of the hive were deposited by the bees in certain privy-cells, and thence removed at intervals by the scavenging authorities. There is nothing in this notion, absurd as it is, outside the scope of bee-ingenuity; on the contrary, such a crude device would be little likely to commend itself to the hive-people, as it would be ridiculously inadequate to the case. How great must be the problem of the preservation of cleanliness in a hive can only be understood when the whole conditions are considered together, and that from a human standpoint. Putting the figures unwarrantably low, what measure of success could the greatest genius that ever lived among sanitary scientists ever hope to achieve, if he were given the task of keeping in cleanly condition, perfect ventilation, and even temperature, a building where 10,000 individuals were crowded together storey above storey—a building hermetically sealed throughout except for one small opening at the lowest level, which must serve for all purposes of entrance and exit to its denizens, as well as sole conduit for the removal of the foul air and introduction of the pure? The task would be gigantic enough in the summer-time, when a large proportion of the inhabitants were away at work during a greater part of the day; but in winter, when all were continuously at home for weeks together, what conceivable device, or combination of devices, could prevent the building soon developing into first a quagmire and then a charnel-house, to which the Black Hole of Calcutta would be a model sanitary retreat?

Yet the difference between such a building and a bee-hive is only one of degree. The same conditions are involved, and the same evils must be combated. Relatively, the problem is the same in each. In the case of the bee-hive, the necessity for this close system of life has been very gradually imposed on its inhabitants; and age-long custom, working on the individual, has at length produced a race marvellously adapted to its special needs. Probably the habit of retention of fæces while in the hive was at first a voluntary one. This, carried on from generation to generation, would react on the physical organism until use became second nature, and finally the present condition was reached. It is a fact that the bee is now incapable of voiding its excreta within the hive, or when at rest. The muscles involved can come into action only during, or immediately after, vigorous flight. In the winter, when long spells of cold occur, not a bee leaves the hive perhaps for weeks together; but an hour's warm sunshine will infallibly bring the whole company out in a little eddying crowd about the hive, and then the necessary action of nature can readily be seen. These cleansing flights occur on all practicable occasions, and fulfil a double purpose; for when the cluster forms again, it will be between combs where the stores are unexploited, and the old, steady, upward feeding-march begins again in a new place. In extraordinary seasons, when the cold weather is much prolonged, the population of a hive may die of starvation within reach of plenty, no opportunity for these flights having presented itself, and the cluster therefore not having left its original station. And here the bee is plainly the victim of her own advanced acumen. Instinct would never have led her into such a foolish plight; but reason, being liable to err, errs here egregiously.

The comparison of a modern bee-hive with a building similar in construction, and as densely crowded with human beings, brings the whole problem to a sharp definition. In such a building, unless a through-current of air could be established, the preservation of life must soon become impossible. Yet the bees have triumphantly overcome all difficulties. Whether in winter or summer, the air within the hive is almost as pure as that in the open, while the temperature can be regulated at will. For the ordinary purposes of the hive—honey-brewing, and the hatching of the young brood—it is kept uniformly at 80° to 85°. When the wax-makers are at work, it rises suddenly to 95° or so, while at the time of the swarming-fever it is often allowed to go even higher. In the hottest days of summer, however, unless the emigration-furore possesses the colony, the interior of a well-made hive seldom shows a temperature of more than 80°. And all this is brought about in a very simple fashion.

The sanitary expert, of merely human stock, could attack the problem in only one way. He must have a through-current of air, impelled either mechanically or automatically; and he must have heating-apparatus acting within the building itself, or warming the incoming draught of air. But the bees work on totally different principles. They will have nothing to do with the through-current system of ventilation. If the ingenious bee-master pierce air-holes in the walls of a hive, the bees will spend the night in carefully stopping them up again. In the old bee-garden we saw the fanning-army drawing out the impure air. These bees had their heads pointing towards the entrance; but, inside the hive, there was another army of fanners, facing the opposite way, and thus helping to drive the same sidelong current. Throughout nearly the whole interior of the hive on hot days fanning-bees can be seen, all helping to keep up this movement. The result is that the pure air, being sucked in at one side of the entrance, flows round the hive and travels out at the other side, much as a rope goes over a pulley-block. The swiftest current of air keeps to the walls and roof of the hive, the air in the centre being changed more slowly. Thus the honeycombs, which are always in the upper stories, lie in the full stream, and the moisture, which the maturing honey is continually giving off, is carried rapidly away; while the brood-combs, lying in the lower, central part, are ventilated more slowly, the air being thoroughly warmed before it reaches them. The larger the fanning-army is, the more swiftly flows the air, and the faster the heat of the hive is carried off. In this way the bees can regulate the hive-temperature to the requirements of the moment, putting more numerous gangs to work in the hottest season, or stopping the fanners altogether in mid-winter, when the natural, buoyant heat-exhalation from the cluster is sufficient to keep up the gentle circulation which then is alone needed.

Sometimes, when the colony is unusually large, the fanning-party will be divided into two detachments, one at each side of the entrance, leaving the centre for the inflow of air. In this case a double-loop system of ventfiation appears to be formed.

CHAPTER VII

THE GENESIS OF THE QUEEN

IT has been said that the ways of the honey-bee are nearly all subject to variation—that in bee-life there are few hard-and-fast, undeviating laws. The rule, of one queen-mother only to each hive, appears to be more absolute than any other, yet it is not without its exceptions. Well authenticated instances are on record where two queens have existed amicably in the same hive, each laying her daily quota of eggs unmolested by the other, and, apparently, with the full approval of the hive-authorities.

It is now also certain that a skilful bee-master can accustom his bees to the presence of more than one queen. Recent experiments in America on this head, although convincing enough as far as they go, need the test of time before their practical value to apiculture can be rightly estimated. To multiply its domestic deities may prove anything but a blessing to the harmony and welfare of a hive. But the fact has been well established that the old rule, of one queen at a time, may be upset—whether permanently, and for the ultimate advantage of honey-making, time alone can tell.

A single queen, when young and vigorous and of good blood, is able to keep an entire hive filled with brood throughout the short honey-gathering season. The brood-nest of a modern frame-bar hive has a comb-surface of over 2,000 square inches, giving about 50,000 cells available for the breeding of young worker-bees. This represents, at times of greatest prosperity, an enormous floating population; but if several queens can be permanently established in one hive, and the hives enlarged to permit each her fullest scope, the figures will soon begin to stretch out into infinity.

Drone-Brood and Worker-Brood

Two facts are well known to experienced bee-keepers—that a large stock gathers more honey than two small stocks containing between them the same number of individuals; and that, when the honey-crop is in full yield, there are seldom enough bees to harvest it. The whole art of latter-day bee-keeping consists in bringing up the numerical strength of each colony to its fullest in time for the great main nectar-flow. Yet, in a good district and in a good season, when huge areas of clover or sainfoin come into full blossom at the same time, and the nectar must be gathered or lost within the space of a fortnight or so, the most populous apiary is seldom equal to the task. Probably, in exceptional seasons, half the English honey-crop is lost for want of bees to gather it. If, therefore, the new system of plurality of queens both justifies and establishes itself, the near future may see a revolution in all ideas relating to bee-manship. All that can be said for certain at present is that as many as five queens have been induced to occupy the same hive in peace and quiet together; but whether this portentous state of affairs can remain a lasting one is still to be proved.

A curious and, to the expert, a startling outcome of these efforts to break down an old and almost universal custom in bee-life, is that the successful establishment of several mother-bees in a single hive appears to lessen the swarming impulse. Hives so treated do not send out a swarm so far as is known. One of the most disappointing experiences in bee-craft is to see prosperous stocks breaking themselves up into several hopelessly weak detachments just before the great honey-flow, when strength of numbers is the one vital thing; and if plurality of queens will prevent this vexatious evil, the old time-honoured custom is sure to go.

The student of bee-life, watching the year's work in the hive from its earliest beginnings, and marking its steady, cautious development, will readily see how the ancient idea of the mother-bee's absolute monarchy gained its vogue. The deception of appearances is all but complete. Right in the heart of the winter-cluster he sees the queen bestirring herself to lay the first eggs, and the bees around her slowly awakening to the duty before them. With the passing of the weeks, he sees the brood-area steadily enlarging; the hitherto close-packed throng of workers gradually extending itself over a larger space of comb; the water-fetchers increasingly busy; the pollen-gathering bees already at work in the crocus-borders of the garden, where the year's first gold and white and purple is gaily flaunting in the sun. He notes that the progress of the colony within the warm hive does not go by the calendar, but checks with each return of cold, and forges ahead only when the spring seems to be coming in right good earnest. He sees, even now, when February is waning and the hazel-catkins fill the bare woodland with a shimmer of emerald, that the colony still husbands its stores, eking them out with a long-sighted parsimony that shall be more than justified when the inevitable cold break comes in the flowery midst of the English May. It is impossible to overlook the evidence of a wise, directing mind through it all; and where should this be seated but in the brain of the single large bee, courted and fed and groomed unceasingly by the attendant host around her—she who is the teeming mother of past tens of thousands, and who carries in her body the seed of all the generations to come?

Yet the truth is that the queen-bee is the very reverse of a monarch, both by nature and inclination. She possesses only the merest rudiments of intelligence. She has a magnificent body, great docility, certain almost unrestrainable impulses and passions, a yielding, womanish love of the yoke; but she is incapable of action other than that arising from her bodily promptings. Her brain is much smaller than that of the worker. In a dozen different ways she is inferior to the common worker-bees, who rule her absolutely, mapping out her entire daily life and using her for the good of the colony, just as a delicate, costly piece of mechanism is used by human craftsmen to produce some necessary article of trade.

In a word, the queen is the sole surviving representative of the aboriginal female honey-bee. The aborted females, the workers, are almost as much a product of civilisation as the human race itself.

Every step of the way now, in a study of the life of the bee, is hedged about with wonders. It is seen that the common worker-bee is raised in a cell allowing her only the barest minimum of space for development, while the queen has an apartment twice as long as she can possibly need. The worker-cells are so designed that as many as possible may be contained in a given area, and their construction involve the least possible amount of material. Therefore these cells are made in the form of a hexagon, this being the only shape approaching the cylindrical—the ideal form—of which a number will fit together over a plane surface without leaving useless spaces in between. Moreover, the cells needing to be closed at the bottom, half the material required for this purpose is saved by the device of placing the sheets of combined hexagons back to back, so that one base will serve for two cells. But it is not only in the construction of the cradles of the worker-bees that rigid economy is practised. From the moment that the egg hatches until the young grub changes into the chrysalis state, it is given only the smallest quantity of food that will support life and allow necessary development.

In the case of the young queen-larva, however, a very different policy is instituted from the beginning. Not only is she given nursery-quarters allowing every facility for growth, but she is loaded with a specially rich kind of food night and day, until she actually swims in it. The nurse-bees are constantly pouring this glistening white substance into the cell for the whole five days of her larval existence, and the effect of this generous diet is obvious from the first in her more rapid growth, as compared with the worker-bee. A further advantage still is that the young queen has perfectly free access to the air at all stages of her development. The worker-cell is but sparsely ventilated, and that only through the narrow top all its six sides and base being absolutely impervious. But the cradle-cell of the queen is not only made of a porous material throughout, but it is commonly placed at the edge of the comb, where it stands out in the full current of ventilation, the air percolating the whole substance of its walls in addition to entering freely at the large cell-mouth. Thus the main cause of the extraordinary difference in the development of the queen-bee and the worker is that of treatment; the one being given unlimited rich food and oxygen and room to grow in, the other receiving only meagre workhouse diet, restricted quarters, and little air to breathe.

Yet, making every allowance for the stimulating or retarding effect of these agencies on the young female grub, we are still hardly any nearer to a solution of the mystery. We are compelled to believe that the egg which produces the worker is identical in its nature with that from which is evolved the queen-bee, because a simple experiment will at once dispel all doubt on the matter. If the egg deposited in the queen-cell be removed, and an egg taken from any one of the thousands of worker-cells in a hive be put in its place, the worker-egg will always produce a fully developed and accoutred queen-bee. On the other hand, if an egg be taken from a queen-cell and placed in a worker-cell, it will as infallibly hatch out into a common undersized worker. It would be sufficient tax on the credibility if the differences of queen and worker were only those of degree. If the queen were nothing but a large-sized worker-bee, in whom certain organs—which were atrophied in the worker—had received their full development, it would be a fact within comprehension; but the queen differs from the worker not only in size and the capability of her organism, but also on several important points of structure. And how can mere food and air and circumstance produce structural change? The worker has many bodily appliances, special members ingeniously adapted to her daily tasks, of which the queen is wholly destitute; while the physical organism of the queen varies from that of the worker in several important degrees.

Some of these must be enumerated. The abdomen of the worker is comparatively short and rounded: that of the queen is larger and longer, and comes to a fairly sharp point. The jaws of the queen are notched on their inner cutting edge: the worker's jaws are smooth like the edge of a knife. The tongue of the worker has a spatula at its extremity, and is furnished with sensitive hairs: the tongue of the queen is shorter, the spatula is smaller, while the hairs show greater length. The worker-bee has a complicated system of wax-secreting discs under the horny plates of her abdomen: in the queen these are absent, nor can the most elementary trace of them be discovered. In their nerve-systems the two show difference, the queen possessing only four abdominal ganglia, while the worker has five. The queen's sting is curved, and longer than the worker's: the sting of the worker-bee is perfectly straight. On their hind-legs the workers have a curious contrivance which bee-keepers have named the pollen-basket. It is a hollowing of the thigh, the cavity being surrounded with stiff hairs; and within this the pollen is packed and carried home to the hive. In the queen both the cavity and the hairs are absent. Her colour also is generally different from that of the worker-bee, her legs, in particular, being a much redder brown.

Here is a problem for our great biologists—a problem, however, at which the plain, every-day man may well flinch. For we seem to have come face to face with new principles of organic life, facts incompatible with the accepted ideas of the inevitable relation between cause and effect. The irresistible tendency at this stage is to hark back; to repeat the experiment of the transposed eggs, and see whether no vital, initial circumstance has been overlooked. But the result is always the same. Nor can the most careful microscopical dissection of the eggs themselves reveal any differences. In this mystery of the structural variance between queen and worker, it would seem that we are forced to accept one of three alternatives. Either the egg contains two distinct germs of life, one developing only under the stress of hard times, the other only to the call of luxury. Or we must go back to mediæval notions, and believe that the worker-bees give or withhold some vital principle of their own during nurturing operations. Or we must give up the problem, and decide that creation works on lines very different from those on which we have hitherto grounded our faith.

The difficulty is further complicated by the fact that this change of nature does not take place until relatively late in the life of the bee. The egg is three days in hatching. But the young larva is at least three more days old before nature has made the irrevocable step along either of the divergent ways. For the experiment of transposition can be made with exactly the same result if undertaken with female bee-larvæ not more than three days old, instead of the unhatched eggs. Indeed, this is an operation that the nurse-bees themselves perform, on occasion. If a hive loses its queen, and it happens that all the eggs in the worker-cells are hatched out, the bees will breed another queen from any one of the worker-larvæ available. This is generally successful when the young grub has not passed the three days' limit. But, even when all the larvæ of the hive are older than this, the bees will still attempt the task, knowing well that, without a queen, the colony must perish. In this case, however, the resulting queen will be defective in various ways. Probably she will never be capable of fertilisation, and therefore the breed of worker-bees will be cut off at its source. Unless the bee-master supplies the colony with a new queen, properly fecundated, the hive will gradually fill up with drones, the old worker-bees will die off, and the stock must ultimately become extinct.

When once the study of the inner life of the honey-bee has been undertaken, the watcher will soon realise that he has embarked on a stranger voyage than he ever contemplated, even in his most daring moments. In the old bee-garden there was a serenity, a quiet enduring bliss of ignorance, that chimed in well with his slothful, holiday mood. The sunshine, the flowers, the song of the wind in the tree-tops, and the drowsy song of the hives; the voice of the old white-headed cottager weaving in his listener's ear the old, comfortable arabesque of error; the sudden, jubilant uproar of a swarm, filling the blue sky with music and the flash of unnumbered wings; the night-quiet, with its deep underground bee-murmur, its dim half-moon peering over the hill-top, the shadowy bent figure of the old beeman listening at hive-doors for the battle-cry of rival queens, that should mean trouble on the morrow—it all comes back to the watcher now as a haven he has left inconsiderately, for a voyage over unknown, stormy seas. For now, with the inner life of the hive going on unmasked before his very eyes, wonder succeeds wonder almost without a break; and each new fact that reveals itself is more perturbing, because more destructive of old, hallowed convention, than any that has gone before.

The hive that has lost its mother-bee, and failed to provide her with a fully developed, fertile successor, is seen to be rapidly declining in its worker-population, while the horde of drones is increasing at a greater rate than ever. But where do these drones come from, if the very fount of bee-life has been dried up at its source by the loss of a fertilised queen? The question brings the student to what is perhaps the most remarkable fact in the whole great book of natural history.

We are not concerned, for the moment, with theological matters; nor will the thread of the story of the honey-bee be laid down, however briefly, for an excursion into the pulpit. Yet here is something that may well give wherewithal for thought. For nearly two thousand years the Doctrine of the Virgin Birth has been the centre of a bitter human controversy. Its liegemen uphold it as a main article of faith, eternally exalted from the odious need of proof; its temperate opposers sadly and quietly set it aside as a natural impossibility. On one side the charge is want of faith; on the other of blind credulity. And yet no one seems to have thought of looking into paths of creation other than human, to see if no parallel exists that may help both sides, and send the swords to sheath before a common mystery. The honey-bee is small among the fowls, but here she looms large in the world, a portentous symbol. It is a fact, now incontestably proved, that the virgin queen-bee is capable of reproducing her kind, yet only the male of the species. If she is born late in the year, when no drones exist, and her fertilisation is therefore impossible, or if some imperfection of wing prevents her going out for her mating-flight, she will still set busily to work at her one function of egg-laying; and these eggs will all hatch out into male bees. The same thing occurs in the case of the queenless hive, which, having neither worker-egg nor worker-grub, whose age is under the three days' limit, yet tries to raise a new queen from a larva perhaps four or even five days old. The queen thus created is queen only in name. She may have her ovaries completely developed, but otherwise she will be congenitally destitute. She will have neither the will nor power to receive the drone; and the eggs that she lays so industriously only add to the crowd of useless males that will soon be the sole representatives of the doomed household.

Following the progress of a bee-colony through the mounting days of spring, we see, with every week that passes, a larger area of comb occupied by the young worker-brood; while about the middle of April the queen pays her first visit to the drone-combs, laying a single egg in each cell, as with the rest. It is commonly supposed that the queen is always surrounded by an adulatory retinue, each attendant bee keeping her head respectfully towards her sovereign, and backing before her as she progresses over the combs. Something of this sort is constantly seen during breeding-time, but at other seasons the queen ordinarily receives little attention, passing to and fro in the hive with no more ceremony than is bestowed on any other of the bees. The mediæval writers were aware that the queen had these attendants, and believed them always to be twelve in number, representing the twelve Apostles. A little observation, however, will soon make it clear that the bees which surround the queen on her egg-laying journeys are neither devotees nor courtiers. They are actually her guides, her keepers. The queen's movements are all prompted by the incessant strokings and pushings and gentle touches of the antennæ that she receives from these. Thus they allow her free passage over the combs, but stop her at each vacant cell, gathering close about her, evidently with the most absorbing anxiety and interest in the operation. First, she peers into the cell, examining it carefully. Then she rears; the bees give way before her; she takes a step or two onward until the end of her body is over the cell. And then she thrusts her abdomen deep into it, pauses a moment, mounts again upon the comb, and the attendant bees at once resume charge of her, and manœuvre her towards the next empty cell.

This process never seems hurried, and yet in the height of the breeding season it must go on at an extraordinary pace. It is well attested that a good queen will thus furnish as many as two thousand to three thousand cells in a day, which gives an average of two eggs a minute, even supposing her to keep at the work without pause for the whole twenty-four hours.

The cells designed to contain the worker-brood measure one-fifth of an inch across the mouth; drone cells are larger, having a diameter of a quarter-inch, as well as greater depth. The queen may pass from one species of comb to the other, but she seldom makes a mistake. The egg deposited in the worker-cell hatches out a female; that which is laid in the larger cell becomes a drone, or male bee. Obviously the deposition of the different kinds of eggs is well under the control of the queen. It will be also seen that not only does the mother bee lay either male or female eggs at will, but their number also is subject to her discrimination. From the time when she begins ovipositing, until she reaches her period of greatest activity in early summer, the increase of the colony is not regular, but goes by fits and starts according to the weather, or the amount of incoming food. If the new honey is steadily mounting up in the storehouse, and pollen is plentiful, the work of brood-raising will go freely ahead; but if unseasonable cold stops the work of the foragers, this will immediately affect the output of the queen, and under exceptionally adverse conditions egg-laying may be entirely arrested. This may also take place in the height of the season, and in full favour of sunshine and plenty, if the hive is a small one, and the limit of its capacity has been reached. The combs will then be full of either honey or brood, and the queen must wait until laying space can be cleared for her. That she is able to do this—that her powers can be augmented or restrained, according to the needs of the colony, and that the proportion of the sexes in the hive can be varied at will to suit like contingencies—can only be understood when the details of her life-history have been passed under review.

In the normal, prosperous colony, which we are now studying, the queen will be in her prime, and under natural conditions will remain at the head of affairs until she goes out with the first swarm in May or June. A queen-bee is at the zenith of her fecundity in the second year of her life. After that, her egg-laying powers steadily decline, although she may live to be four, or even five, years old. But the authorities in a hive rarely allow a mother-bee to retain her position after she has shown signs of waning energy. Preparations are at once set on foot for the raising of another queen.

Queen Bee in Breeding-Season.

*The Queen is Seen in the Act of Laying,
with her Circle of Guides and Keepers about her*

A very old queen will have lost her power to lay worker-eggs, and will have become nothing but a drone-breeder. But the bees are seldom caught napping in this way. Long before this happens the building of the royal cells will have commenced in the hive. A queen-cell has been likened, by various writers, to an acorn, and when half completed it bears a very close resemblance, both in size and shape, to an inverted acorn-cup. This is commonly hung mouth downwards at the side or base of one of the central brood-combs, but it may be placed right in the middle of the comb, in which case the cells around it are cut away to give it air and space. Whether the old queen herself deposits the egg in the royal cell—thus unwittingly supplying the means for her own future dethronement—or whether the worker-bees transfer to it an egg or grub from a common cell, is not yet finally ascertained. As, however, the mere sight of a royal cell usually excites the queen to fury, the chances are that she is never allowed to approach it at any time, and the egg would then be placed there by the worker-bees. But, in the great majority of cases, it is probable that new queens are raised by enlarging an already existing worker-cell, in which an egg has been previously deposited. As far as is known, this is always the case when a young grub is used for the purpose instead of an egg. It is possible, also, that the queen is physically incapable of laying in a royal cell an egg that will produce a female bee; but this curious point will be touched upon at a later stage.

The old trite saying among bee-men, that bees never do anything invariably, receives constant illustration in any near study of the ways of the honey-bee. It has been seen that a colony deprived of its queen, and having no worker-egg or grub less than three days old wherewith to make good its deficiency, is commonly doomed to early extinction. But, on rare occasions, colonies supposed to be in this plight will make an unexpected and inexplicable recovery. After a period of the doldrums, extending for three weeks or more, a sudden renewed activity and exhilaration is observable in the hive. The pollen-bearers, who have been hitherto almost idle, resume their busy work; and, on the hive being opened, all the evidences of the presence of a fertile, laying queen-mother are again to be seen.

In many instances in which a new lease of life has thus been vouchsafed to a colony under what seems an inexorable ban, no doubt appearances have been deceptive. The bees may have discovered in their midst a worker-larva not yet too far advanced for promotion to queenship, and thus have achieved their salvation at the eleventh hour. But, in at least one case, the testimony against the possibility of this seems complete.

A Queen-Cell

A nucleus stock, containing only three or four small combs and only about five hundred bees, was deprived of its queen. Ten days later every queen-cell that had been formed in the interval was destroyed, leaving in the hive not a single egg or bee in the larval state. Nevertheless, on the hive being opened after a further period of eighteen days, one new queen-cell containing an egg was discovered. And this egg duly hatched out into a fine, well-developed queen-bee. Assuming the facts to be true, and they seem to be incontrovertible, there is only one inference to be drawn from this: some enterprising bee of the colony must have gone to another hive and either begged, borrowed, or stolen a worker-egg. Apiarian scientists very rightly hesitate to ascribe to the honey-bee surpassing ingenuity of this kind on the testimony of a single case, however well authenticated. But other instances are on record nearly as indubitable, and as it is an unquestioned fact that worker-bees will carry eggs about from comb to comb within the space of their own hive, it does not seem wholly incredible that they may visit other hives in the immediate neighbourhood, especially when impelled to extra resourcefulness by so vital a need. The whole question is interesting in more ways than one, as it seems to bear very trenchantly on the problem of "Reason *versus* Instinct," now busy in the thoughts of most modern naturalists.

In whatever way the egg for the queen-cell may be furnished by the stock intending to raise a new mother-bee, the first sign of life is always the same—a tiny, white, elongated speck, glued on end to the base, or what must rather be called the roof, of the inverted cell-cup. In this state it remains about three days, when the larva hatches out, and at once the special treatment accorded to the young queen begins. She is loaded with rich provender from the first moment of her existence, living literally up to the eyes in the white, shining, jelly-like substance that the nurse-bees are continually regurgitating and pouring into the cells. This superfeeding process is continued for about five days, when the larva has reached its full growth and the cell its greatest dimensions. The larva then stops feeding to spin itself a silken shroud before changing into the pupa state, and the bees seal up the door of the cell. In its completed state the cell loses its resemblance to an acorn, and is rather to be likened to a fir-cone. In the case of the common workers and drones, the cells are made of pure wax, only the capping being of mingled wax and pollen; but the queen cell is constructed throughout of this porous material.

The fully grown queen-bee is ready, and more than anxious to leave her cradle-cell in about fifteen or sixteen days after the laying of the egg. The bees, however, generally give her a first lesson in obedience even at this early point in her career. It is a critical time in the history of the hive, and much thought and care have been bestowed on the complicated business in hand. In the first place, it would never have done to allow the whole future welfare of the colony to depend on a single life alone. Therefore not one queen has been raised, but several. As many as five or six queens may be ready to hatch out in different parts of the brood-nest, and none of them will be permitted to break from her cell until the appointed time arrives. For each the cradle now becomes a prison. A small hole is bored in the cell-wall, through which the impatient captive is fed, pending the day when she is to be allowed her liberty; and close guard and watch is kept over each cell to save it from the violence of the old queen, who is becoming hourly more restless and suspicious.

The complete subjection of the mother-bee to the ruling worker-class in the hive receives here a striking confirmation. She is a true exemplar of a prevailing kind of femininity—comely of person, untutored in mind, an inveterate stay-at-home, a prolific mother; and now there awakens in her the sounding chord of jealousy. Left free to act on her own impulses, she would soon bring about a speedy end to all the careful, long-sighted preparations within the hive. She would tear open each royal cell, and with one thrust of the curved, cruel scimitar that queen-bees use only on their equals in rank, its occupant would be ruthlessly despatched, and her own supremacy reinstated. But an impassable barrier stops the way—the collective will of the hive. The violent delight of killing has once been hers; she will never know it again. Now her own fate is in the balance. It may be death, or a new life in a new home: all depends on the deliberate decree of those who have made her, and who now use or discard her, for their own purposes. If it be late spring, and the condition of the stock warrant it, this governing spirit may decide for colonisation, and the old queen may be disposed of by sending her off with a swarm. But other counsels may prevail. The times may be unripe, or the weather inopportune. And then Fate, in the shape of a merciless application of principles, will descend upon her, and her own wise children will ruthlessly put her to death.

This State-execution of the queen, at the first sign of waning fertility, is a peculiarly pathetic as well as a tragic phase of bee-life. The stern, soured amazons of the hive must have their systems and conventions in everything they undertake; and they cannot even bring about the supersession of the old queen without due circumstance and ceremonial. Given that it would be against the best interests of the common weal that she should retain her life after the loss of her queenhood, one swift stroke would immediately determine the matter, and the law—that there shall be no useless members in the bee-republic—would have its due fulfilment. But old tradition rules that the queen shall suffer no violence from the weapons of the common herd. She is to die, but her death must be brought about in another way. And so the fawning executioners gather round her, locking her in an embrace that tightens with every moment, until the breath is literally hugged out of her body. All her life has been spent in the midst of caresses, and now she is to die of them, close held to the last in that silent, terrible grip.

CHAPTER VIII

THE BRIDE-WIDOW

IN the heat and glow of the fine June morning you may see her, the young virgin queen, making ready for her nuptial flight.

At first she is all hesitancy; wandering to and fro amidst the crowd on the hive-threshold; coquetting with the sunshine; loath to return to the dim, pent, murmurous twilight she has forsaken, yet hardly daring to launch herself on wings that are still untried.

For three long days and nights since her release from the prison-cell she has been a curiously solitary figure in the busy throng within the hive. Instead of the enthusiastic, welcoming world she expected, she finds none but unregarding strangers about her. Not a drone glances her way, and the worker-bees go upon their business in seeming unconcern at her presence. They do not even trouble themselves to feed her, and she is left to forage for herself as best she may. A conspiracy of indifference is on the clan—all part of a deep design for her education, if she only knew it, but singularly damping to the ardours, and great ideas of destiny, that gather within her day by day. At length the call comes for which all are secretly waiting, and obeying irresistibly, she presses out into the light.

As she stands hesitating, the hot June sun falls upon her, laving her in molten gold. The blue sky beckons her upward. All the world of colour and incense and life calls her to her wooing, and she must needs obey. With a little glad flutter of the wings, she breaks at last from the scrambling company about her, and soars up into the light.

Warily now she hovers, taking careful stock of her home and its surroundings. Then round and round, in ever widening and lifting circles, each sweep upward giving her a broader view of the world that lies beyond. And then away into the blue sky so swiftly that no human eye can follow; yet only for a short flight. She is back again now, almost before you have missed her, and hurrying, frightened at her own audacity, into the old safe gloom of the hive.

Thus she dallies, to and fro between the sunshine and the darkness, each time adventuring a little farther into the blue playground of the upper air, until at length the inevitable comes to pass. A great drone—one of the roistering crowd that fills the bee-garden with its hoarse noontide music—spies her, and gives instant chase. At sight of him she wheels, and darts away into the sunshine at lightning speed. Yet the first drone has hardly stretched a wing before another is after him, and still another. Thick and fast from all points they gather for the race, until the fleeing queen has drawn a whole bevy of them, streaming like a little grey cloud behind her. This much you can see as you strain your eyes in their track; but in a moment quarry and huntsmen have vanished together, volleying, as it seemed, straight up into the farthermost skies.

From her birth to the day when that terrible, living cordon closes about her, almost the whole life of the queen-bee can be followed step by step. Only this one moment of her bridal stands unrevealed, and perhaps for ever unrevealable, to human eyes. You can picture to yourself the wild chevy-chase through the clear June air and sunshine; you can give, in fancy, the prize to the strongest and the fleetest; but all you will know for certain is that in a little while the queen returns to the hive, sobered and solitary, trailing behind her infallible evidence of her impregnation and the death of the victorious drone. She has been the bride of a moment; now she is to be the widow of a lifetime. Henceforward her days are to be spent in the twilight cloisters of the hive, flying abroad so rarely that many an old experienced bee-man will say she comes forth only once a year when she leads a swarm. But in her body now she carries the seed from which will spring up a whole nation. Before her marriage-flight she was the least considered of all the colony; now she is welcomed home with public ovation; lauded, fed, and fondled; set up in the high place, a living symbol of the tens of thousands unborn. As in olden, savage times, the royal festivals had their human sacrifices, so this paramount day in the perfected communism of the bee-people must vent its rejoicing in slaughter. But it is not tribute of common slaves that is now to redden the State-shambles, nor will the work fall to the common executioner's knife. There are captive queens in the citadel—a royal sacrifice ready to hand, and a royal blade hungering for the task. Once the queen has proved her intrinsic motherhood, and the first few worker-eggs have been laid in the comb, the guards will stand away from the royal prison-cells and let her wreak her will upon them. It is all very ghastly in a miniature way, yet very queenly, as old traditions of human queenhood go. She gives over her nursery-work gladly enough for a moment, and flies to the slaughter, tearing down the prison-doors, and putting each clamorous captive fiercely to the sword.

Apart from this tragic element of sororicide, quickly over and soon forgotten in the general rejoicing, there is true romance in the early life-story of the Queen of the Bees—bridehood, wifehood, widowhood, following hard upon each other, all in the space of a single hour. But in the details of her common everyday life that succeed this tense period, above all in the wonderful structure of her body and its functions, there is greater romance still. That she has but a single commerce with the drone, and thereafter is exalted to perpetual fecundity; that, through her, sons and daughters can be given to the hive in just the proportion needed for the good of the State, or that increase of population can be wholly arrested at will, are facts to be accredited only after sure knowledge. And to understand how these results are brought about, it is necessary to learn something of the anatomy, as well as the manner of fecundation, of the mother-bee.

In the first place, as fertilisation of the one sex by the other is usually regarded, the queen-bee is not fertilised at all. The vital essence of the drone does not penetrate the ovaries of the queen, but passes immediately after coition into a receptacle specially provided for it, where it is stored, and its effectiveness preserved, during nearly the whole lifetime of the queen. It has been shown that the virgin queen is able to lay eggs from which only drones, or male bees, originate. The fecundated queen, however, can lay both male and female eggs, and she has the power of depositing either kind when and wherever she wills. The whole thing, amazing as it is, and far-reaching in its results, has, like many other extraordinary devices in nature, a simple explanation. The gland wherein is stored the male life-essence can be opened or closed at the will of the mother-bee, or rather, as will be shown, according to circumstances that for the moment involuntarily but inexorably guide her. When she is brought to the large drone-cell, this gland remains shut, and the egg escapes without contact with its contents. But at the narrow worker-cells the gland in the oviduct is opened, and the egg, in passing, absorbs some of its containing germs. Thus only the female bee is born of the union of the two parents; the male bee is the offspring of mother alone.

Of this primal incident, the parthenogenesis, or birth of the fully equipped male from the virgin female, little more can be said than that it is a well-ascertained fact of nature, exemplified in several other insects beside the honey-bee. But while we are witnessing the part played in the hive by the fecundated queen, with her elaborate organism, much is to be noted; and here we really get the master-key to a right understanding of the whole system of bee-government. It would be an anomaly if the highest, most important functions of the State had been entrusted solely to the queen, whose feeble intelligence renders her, of all others, least likely to execute them properly; and we find, in fact, that no such reliance is placed on her. The worker-bees, who take her in charge on her return from her mating-flight, henceforth originate her every act and impulse. It has already been seen how she is led from cell to cell over the combs; how she is caused to lay, in earliest spring, only a few eggs a day, while in the summer she may produce several thousand; and how her output may be checked or augmented at any point between. Now we are to realise how it is all brought about; or, at least, bring conjecture as near to certainty as may be with so difficult a theme.

During the first two days of her life as a perfect insect, we saw the young virgin queen mingling with the throng in the hive almost unnoticed, and left to seek her own food from the common store like the rest. But now that her fecundation has been achieved, she has a whole suite of chamber-women, whose principal duty is to attend to her nourishment. From their mouths they feed her, giving her, in all probability, the same rich substance that was administered to her when but a larva in the cell. This bee-milk consists mainly of honey and pollen pre-digested, but it has been proved that its composition can be altered at will by the ministering bees. Additions to it are made, either separately, or combined in varying proportions, from three or four distinct glands, each of which exudes a liquid differing in nature from that of the rest. The particular kind of nourishment given to a queen who is to be urged on in the work of egg-laying, has the effect of stimulating her ovaries. The more food of this kind she receives, the greater will be her prolificacy. On the other hand, a diminishing allowance will mean a corresponding decrease in her egg-laying powers; while, if this rich diet be withheld altogether, and she is forced to help herself from the honey-cells, the development of these eggs may cease entirely, as it actually does in the coldest time of the year. Thus the bees play upon her, producing just the music needed for their purposes. As the days lengthen, and the spring sun gets higher and warmer, they gradually waken her docile nature to its one paramount task. In the flaming weeks of summer she sits at an unending banquet. And when autumn comes, with its chilly nights and steadily failing sun-glow, the generous fare is slowly withdrawn; her retinue thins and disperses; at length she becomes a solitary, unmarked wanderer again, sipping, with the commonest worker, at the plain household sweets.

How the proportion of the sexes is so unerringly regulated by the hive-authorities through their influence on the mother-bee, is not so readily explained; nor can it be at present more than shrewd conjecture, a backward reckoning from effect to cause. Probably the opening or closing of the fertilising gland, which decides the sex of the egg, is automatic, the attitude of the mother-bee during oviposition determining its action. When she enters the narrow worker-cells, her body is necessarily straightened, and this may produce pressure on the fecundating gland, resulting in the impregnation of the egg. But in the wider drone-cell no such constricted posture is needful, and the egg may therefore pass untouched by the fructifying germ. If this version of the matter be accepted, the natural inference is that either the mother-bee is incapable of laying female eggs in the cells specially constructed for raising queens—these being the largest of all,—or that there is something in the peculiar curve of the cell-cup which compels her to straighten her body in the act, and so brings about the same posture as with the narrow worker-cells.

This theory, although at present the most plausible, has received, it is true, little confirmation in fact. No one, apparently, has ever seen the mother-bee lay in a queen-cell, nor has the transportation thither of a worker-egg by the bees actually been witnessed. To cling to the old idea of the supremacy of the queen-bee, giving her the power and ability of a despotic, all-wise sovereign, would, of course, set this and many other vexed questions at rest. Nothing, however marvellous, would be too much to expect of her. But the farther the student of bee-life goes in his absorbing subject, the more impossible the old notion seems. Proof comes to him with every hour that the mother-bee is virtually a servant, and never a ruler in the hive; and just as assured testimony reaches him of the universal potency of the worker-bees. All else that takes place within the hive is brought about by their collective will and agency; and it would be strange indeed if this vital matter of progeneration were not subject to the same controlling force.

CHAPTER IX

THE SOVEREIGN WORKER-BEE

WATCHING the inner life of the hive in the season of its full activity, it is not the untiring spirit of industry pervading the whole bee-commonwealth that most excites the student's wonder, but rather the fact that this ceaseless diligence finds so many outlets—that so many different kinds of necessary work are going forward at one and the same time.

Between the brood-combs the nurses are feeding the young larvæ, or clearing out the empty cells, or sealing over the full-grown nymphs for their pre-natal slumber. Hard by, the sowers are at their vital work, driving their living seed-barrow, the queen, over the combs. Elsewhere the wax-makers hang in a silent, densely packed cluster. Overhead, the new honey-combs are growing; the masons building up the cell-walls, while the engineers devise means and ends, calculate strains, put in a strut here, a stay there, or flying buttress from one comb to another, or cut new passage-ways where the traffic seems too congested for the old thoroughfares of the hive.

On all sides the scavenging bees go to and fro, picking up every particle of refuse, and carrying it safely away. Winged undertakers drive their trade in the midst of the throng, bearing the corpses of their comrades, old and young, towards the entrance, and flying away with them into the sunlight of the young spring day. There is the ventilating army outside the city gates, skilfully organised in relays, so that, day and night, a constant circulation of air is maintained. There are the guard-bees close by, watching all in-comers and out-goers. There is a sort of General Purposes

Committee ready outside the threshold with a helping hand for all: succouring the overladen, grooming down any in need of such assistance, gathering up fallen treasure, or, as it would seem, taking careful note of the weather for their next official report. And all through the hours of sunshine, in unnumbered thousands, the foragers are charging to and fro, some bringing nectar, some staggering in under mighty loads of pollen, others with full water-sacs, still more dragging behind them lumps of the curious cement called by the ancients Propolis, and used for so many different purposes in the daily work of the hive.

And it all goes on with the regularity of a well-ordered human settlement. There is complexity, yet no confusion; there is speed without hurry. Each busy gang of labourers has apparently a distinct and definite task allotted to it by the central hive-authority; co-operation and progress are, to all appearances, deified cause and effect in all the affairs of the hive.

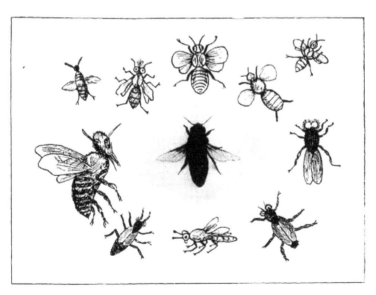

The Honey-Bee (Enlarged)
From Life: and as Some of the Ancient
Draughtsmen Depicted her.

It is easy—nay, inevitable—in any close study of bee-life with the help of the modern observation-hive, to overset the ancient idea of absolute bee-monarchy under a single king or queen. But it is not so easy to determine how the general government of the colony is actually carried on. Innumerable small consultations on minor matters are seen to take place on every side during each moment of the busy day; but nothing like general communication is ever visible. And yet, how are the great national movements, such as the despatch of a swarm or the supersedure of an old queen, brought about; how are the various common crises of the State met, and provided for? The only rational inference seems to be that each worker is in herself the perfect evolved presentment of republicanism, in whom all imaginable difficulties in collective life have their best solution, tried and proved through the ages, and resorted to unerringly as a matter of course. Thus a common need is felt, and met instantaneously by a common, recognised expedient. The judgment of one is necessarily the judgment of all. Every problem of daily life, however intricate, is solved by the one device, brought to the fine point of perfection through the experience of countless generations, and applied by each individual to the common want, just as hunger impels all mankind to eat.

Such a condition of affairs, even in a community of human beings, would imply a very high state of mental, if not of moral, development in the individual. It would mean entire negation of self in the interest of the common good. Even with all the forces of heredity at work, it would need stern ascetic training for the young, and for the transgressing adult a swift and merciless retribution, if the last dream of communism—the abolition of all law and penalty, and the establishment of a natural autonomy of well-doing—were ever to be realised in fact. And yet some such state of things appears to exist in the bee-commonwealth: the individual worker-bee seems to be the product of some such system carried on through an indefinite space of time. Order is preserved, public works go diligently forward, the clock of the national progress keeps time to the second, not because there is a central wisdom-force to plan, to govern, to awe recalcitrants, but because every worker-bee is herself the State in miniature, all propensities alien to the pure collective spirit having been long ago bred out of her by the sheer necessities of her case.

The worker-bee, as we see her in the hive today, although evolution must have been busy through the ages determining her present mind-power and bodily conformation, is nevertheless as much a product of direct artifice as she is of original nature. We have seen how the egg containing the feminine germ, if given full scope and opportunity, develops into what may be taken as the complete aboriginal type of female bee, differing from the worker in a dozen essential ways. The queen also is probably, in one respect at least—her amazing fecundity—a deliberate creation of the hive-people, as her over-production is brought about by over-stimulation to meet an artificial state of affairs. Left to herself, under pristine conditions, she would certainly lay on a much more moderate scale. But the worker-bee owes her unique structure and mental constitution almost entirely to the intervention of her nurses from the moment of the hatching of the egg. Careful experiment has proved that the queen-larva and the worker-larva are identical up to the third day of their life in the cell, except that the queen has made more rapid growth owing to more generous and more ample fare. After the third day, the genital system of each larva will begin to develop, if this rich nitrogenous diet is maintained. In the case of the queen, this pre-digested food, well called bee-milk, is lavished on the favoured grub up to the last moment of its larval existence, no other food being given. But in the case of the worker-grub, not only has its supply of bee-milk been restricted in both quantity and quality from the day of its birth, but now—just before the development of the ovaries might be expected—an important change is made. The allowance of bee milk is greatly reduced, while plain honey is given in addition, but on the same parsimonious scale, to the end of its five days' larval life.

What other influences, if any, are brought to bear on the young worker-bee at this portentous stage of her career, it is impossible to say. But at least the change in the food is well ascertained, and the results—whether of this alone, or in combination with other treatment—are more than astounding. Not only is the development of the sex-organs so completely arrested that hardly a trace of them can be discovered in the adult worker-bee, but, from that moment, the larva seems to become an essentially different creature, reflecting more and more the attributes of her nurses, and showing wider and wider departure from those of the mother-bee. As soon as the worker changes into the pupa state, organs appear of which the queen has not the faintest rudiments. She receives her special equipment for field-work in a pair of baskets for carrying pollen. Her tongue is lengthened, so that it may reach the nectar hidden deep down in the clover-bells. She is to become a builder, and therefore is provided with half a dozen crucibles wherein to prepare the wax. Her useless ovipositor is changed into a weapon: it is straightened, shortened; the barbs upon it are multiplied and strengthened; a gland, with which it is furnished, and which, in the queen, contains an all but harmless fluid, is now filled with an active poison. Above all, she develops a brainpower far in excess of that of the normal female bee, her mother; and she acquires a whole new set of impulses and aspirations from beginning to end.

While the queen-bee's natural element is the obscurity of the hive, and she would seem both to hate and fear the sunshine, the worker is essentially an outdoor creature, revelling in the light and air. While the queen, though obedient to the destiny that has made her over-fruitful, displays nevertheless not the slightest joy of motherhood nor interest in her children, the worker, doomed to eternal spinsterhood, yet constitutes herself the true mother and nurse and instructress of all the young in the hive. And the price exacted for the authority and power which she usurps, or was usurped for her by those remote ancestors of hers who first invented the sexless honey-bee, must be paid in the hardest coin—that of life itself. Instead of the years that nature allotted to her kind in the beginning, she is to endure hardly as many months. Destiny, and her own vaulting ambitions, have given her too arduous a part to play. Her stunted, yet over-elaborated body and over-developed brain, cannot long hold out against the wear and tear of the life she is born to. At best a few months see her dead at her work, or using the last pulsations of her worn-out, ragged wings to carry her away to the traditional burial-place of the hive; or her end may be to fall under the stroke of the State executioners. For the old-age problem has long ago discovered its effective solution in the bee-republic. Justice that is capable of being tempered with mercy carries its own mark of imperfection indelibly upon it. When the principle of all for the common good has been driven to its last resort in logic, mercy to the individual can only be another name for robbing Peter to pay Paul. In bee-communism the sole title to life is utility, and so the old worn-out, useless workers must go.

The development of the worker-egg through its various stages of growth, until the perfectly formed insect emerges from the cell, makes a curious study. The egg itself is remarkable, for it is covered with an hexagonal pattern. The large compound eyes of the fully grown bee also show this form. Each eye consists of about four thousand separate lenses, and each lens is a regular hexagon. Wonder has often been expressed at the ingenuity of the comb-builders in making the cells six-sided, and thus crowding into a given space more compartments than could be secured by the same amount of material wrought into any other shape. The ancient writers explained this choice of the hexagonal cell by the supposed fact that the six legs of the bee were simultaneously employed in comb-building, each leg constructing its own portion of the cell. A more modern idea is that the particular shape of the cell is accidental, or rather the outcome of compelling circumstance, mutual pressure causing the cells to assume the hexagonal form.

Now, it is quite true that soaked peas in a bottle will take this shape in swelling, but the analogy will not hold good in respect of comb-building. In the work of the bees there is no pressure or constriction of any kind. Each cell is made separately, being joined on to those above it; and the comb expands steadily downward and sideways through an empty space until the desired limit is reached. A much more probable explanation of the hexagonal form of the cell is that it was arrived at by experience. The first combs may have been built with round cells, the interstices being filled in with wax. But the bee, who is an expert in the science of economy, would quickly see the disadvantage of this plan. And with the hexagonal principle, an old familiar thing in the hive—witness the pattern on the egg-surfaces, and the compound eye-construction—it would not be long before she hit upon the better, more scientific way.

There is, however, another reason, and almost as potent a one, for the adoption of the six-sided cell both for brood-raising and the storing of honey. It must be remembered that the present system of vertical walls parallel and close together, made up of numberless small horizontal chambers placed back to back, is not an ideal arrangement either for the raising of the young or the storing of food. Yet it is the best possible contrivance under the circumstances, which are forced upon the bee by the necessity of leading a close, crowded, communal life. Air is a prime need for all operations in the hive, but for none more than the development of the young bees. When a queen is to be raised, a full supply of fresh air is given her, but only at the expense of valuable space. With the common kind, of which perhaps ten or fifteen thousand may be maturing in the brood-nest at one and the same time, it is obviously impossible to make any such concession. The young worker- or drone-larva must secure what air it can through the narrow cell-top. Now, the bee breathes at all stages of its career not through the mouth, but by means of air-holes or spiracles in the sides of its body. If the cell were round, the larva, when fairly grown, would fill the space, and the air would reach the spiracles only with difficulty. But, no matter what the size of the young grub may be, the angles of the hexagon cell are never quite filled. They form half a dozen by-passes for the air, arranged on all sides, and extending right to the base of the cell; and thus the larva has the full benefit of the available air-supply, even though it be necessarily scanty.

With the store-combs the six angles of the cell fulfil an equally important office. The ideal honey-cell would be one with its mouth opening upwards, so that it could be filled in an ordinary rational way. But under the strict economical principles ruling in the hive such an arrangement would be impracticable. The honey-vats must be stacked one over the other in a horizontal position, and therefore must be chargeable from the end. All cells in the comb have a slight upward tilt, but not enough to retain the fluid contents if the cell were a round one. The effect of the angles in the hexagon is to increase the retentive property of the cell, and experience has taught the bees how to supplement this natural holding power of the angles by just that slight cant of the cell which is necessary to prevent the nectar running out.

The worker-bee, during her period of larval life, at first lies coiled up at the bottom of the cell, but as her size increases she takes up a position lengthways, with her head towards the cell-mouth. This, however, is not a constant attitude, for she seems at intervals to make a series of slow gyrations or somersaults, probably to facilitate the casting of her skin, which she accomplishes several times during her five days' life as a grub. At the end of this time the nurse-bees stop the feeding process and seal up the cell. Now the larva sets to work, first to spin herself a silken shroud before entering on her long sleep as a chrysalis, and then to change her skin for the last time. In the case of the worker these fine-wrought sleeping-clothes envelop her whole body, forming a continuous cocoon. But the queen-larva weaves herself only a scanty sort of cloak, covering her head and thorax, but leaving her nether portions bare. The theory usually advanced in explanation of this is, that when the surplus queens are slaughtered in their cells by the accepted mother-bee after her fertilisation, the fell work is rendered easier by the absence of the tough material of the cocoon over the parts generally attacked. It seems to be well substantiated that in a battle of queens the stings are not used haphazard, as with the workers, but each queen tries to thrust her weapon into one of her enemy's spiracles or breathing-holes, of which she possesses fourteen, seven on each side. And a stroke dealt in this way appears to be always fatal.

But, in all likelihood, the true reason why the queen sleeps in a short gown made of tough, coarse fibre must be looked for somewhere back in the old ancestral history of the honey-bee. It is probably safe to consider the complete worker cocoon as a comparatively recent introduction, evolved to meet some necessity arising since the bee-people became a civilised race. But what its true origin was appears to be out of the reach of all conjecture. A curious fact is that these cocoons are never removed from the cell. They remain fixed to its sides throughout, and though the cell is otherwise carefully cleaned after the young bee has vacated it, the cocoon is never interfered with, but continues as a permanent lining to the cell. The same thing occurs with all successive generations, each bee leaving her swaddling-clothes behind her, until so great an accumulation occurs that the cell becomes too small for breeding any but a puny, undersized race. With wild bees, where the nest has been constructed in a tree-hollow, and there is usually plenty of surplus room, the old brood-combs may be eventually abandoned and fresh ones built farther on. Thus the stock generally shifts its station from year to year. These natural bee-nests, or bee-bikes, as country people call them, often reach a great age. Sometimes a swarm will get under the rafters in a house-roof, and may be left undisturbed for generations. In one case bees were traditionally supposed to have inhabited a blind loft in a farmhouse continuously for forty or fifty years. A legend rife in the village credited them with having stored many tons of honey, but when the stock was sulphured little more than a vast accumulation of comb was discovered. This comb was of all ages, from a few weeks old to an unconjecturable number of years. Much of it was perfectly black, and the cells choked up with pupa cocoons.

The fact that egg-laying is continued in these combs where others are not available, even though the capacity of the cells has been greatly reduced, seems to cast an added doubt on the theory that the size of the cell is responsible for the fertilisation or non-fertilisation of the egg as it is deposited by the queen. Very old drone-comb is sometimes found in use for breeding purposes where the cells have become no larger than those used for normal worker-brood. And yet the queen continues to lay in them unimpregnated eggs. The whole question is still hedged round with difficulties.

The young worker-bee, at the end of about three weeks from its first inception, breaks from its chrysalis-skin, and begins to gnaw its way through the cell-cover. The pollen, which is combined with the wax to form this capping, discharges a double office. It makes the wax porous for the admittance of air, and it renders the cell-cover edible, thus causing the young bee to effect its own release through the promptings of its appetite. The new-born worker, although fully grown, is a weak, greyish-hued, flaccid creature for some time after it leaves its cradle. Its earliest impulse seems to be to groom itself, and then to wander about on a tour of inspection of its as yet narrow world of gloom and noise and bustle. For the first day or two it does little else than crawl about unnoticed in the busy throng, gradually gaining strength and rigidity of limb. On the second day it may be seen dipping into the open honey-vats and pollen-bins, of which a few are always scattered here and there among the brood-cells. After this it seems to waken in earnest to its duties and responsibilities, and takes its place among the nurse-bees, setting to work with the rest in the stupendous task of feeding the larvæ.

In the ordinary course, the young worker-bee will not leave the hive for about a fortnight after its emergence from the cell. In the interval, however, it has a whole policy of life to study, and several trades to learn. All the indoor work of the hive appears to be done by the young bees during these first weeks of their existence.

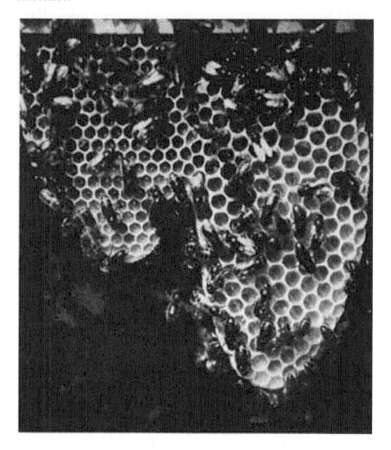

*Brood-Comb,
Showing the Two Sizes of Cell Together,
with Eggs Cemented by Queen to Cell-Bases*

On them the whole care and sustenance of the young brood depend. They produce the wax, and build the combs; they look after the order and cleanliness of the hive; they are the brewers of the honey, and the keepers of the stores; they feed the queen-bee on her ceaseless rounds, and also give the drones their daily rations of bee-milk, for it is certain that the male bees depend very largely on the workers in this way, drawing only a part of their diet from the common stores. The old bees are the foragers; but it is probable they are met by the younger ones soon after their return to the hive, and their burden of nectar, being regurgitated, is transferred to the pouches of the young bees, by whom it is carried to the store-combs in the upper regions of the hive. At least, if the storage-chamber of a hive be opened during the busy part of the day, hardly any old bees will be seen among the crowd, which is industriously filling the cells with the new-gathered sweets.

It is not until the beginning of the second week of their life that the young bees make their first essay in the open air, and then it is only for a few minutes during the hottest part of the day. This sudden midday uproar is a familiar experience to the bee-keeper during the late spring and summer; and although the drones at first contribute largely to the chorus, they soon fly away, while the singing cloud of bees which remains enveloping every hive at this time, is entirely composed of the young house-bees taking their daily brief allowance of exercise and air.

It is found that the glands necessary for the production of the brood-food, as also the wax-generating organs, are largely developed in bees only a few weeks old, while, after their first month of life is over, these organs are greatly reduced. The bee generally begins outdoor work as a forager soon after she has reached the age of fourteen days. It is, however, probably a week or two longer before she attempts the more serious business of nectar-gathering. Nearly all the pollen-bearers are bees in their first young strength and vigour, and therefore peculiarly adapted to the carrying of heavy burdens. But as soon as the worker-bee has settled down to the great paramount task of honey-getting, she seems to leave the pollen alone. Thus, in a normal colony, the life of the honey-bee, short as it is, is carefully planned out from beginning to end, each period having its special task for which the age of the bee is peculiarly fitted. Yet this rule is no more absolute than any other of the ways of the hive. Where the community is short-handed, and there are not enough mature workers to gather stores, the young bees will be turned out to forage at a much earlier date in their career. In the same way, if a hive has been without a queen for some time, and therefore few young bees are available to care for the brood when the new mother-bee has at last established herself, many of the old workers will stay at home and busy themselves with the nursery-work, which in the ordinary course they would have long since relinquished.

There are many such instances of ingenious makeshift, or special adaptation, in the ways of the honey-bee. She is a creature full of resource on emergencies, but it is in the provision of desperate remedies for really desperate ills that she shines at her brightest. The prime disaster in bee-life is the loss of a queen at a time when it is impossible to appoint a successor. The standard of intelligence, as well as that of character, varies among bees almost as much as it does among men. Some colonies will work harder and for longer hours than the rest. Others will ease off when they have put by what they consider a sufficiency of stores, and an idle spirit spreads visibly among them. In a few cases there is a distinct moral twist in the national character, and the bees take to robbing their neighbours' larders instead of working to furnish their own.

Permanent queenlessness is a calamity which affects different colonies in different ways. With some it means complete despair, a cessation of all enterprise or interest in life. Work is stopped; the guards are withdrawn from the gate; the community seems to give up in a body, and to await extinction with no more hope than a batch of criminals in the condemned cell. But with others the common disaster is but a signal for a universal quickening of wits, a furbishing-up of all possible and impossible resources. To bees of this temper we should look for such episodes as the egg-purloining to supply a queen-cell, which has been already dealt with. But for supreme ingenuity, even though it be the forlornest of forlorn hopes, perhaps there is nothing to equal a device sometimes resorted to in this last emergency.

Looking through a hive which is not only without a queen, but which is without any means of raising one, certain mysterious eggs are unexpectedly discovered. These eggs are obviously quite newly laid, but not in the orthodox way. A normal queen works consistently from cell to cell, over a fairly regular patch of comb, and deposits only one egg in each cell; but these eggs in the queenless hive have been laid in a curiously haphazard way. The eggs are straggled over the comb. Two or three cells have been furnished at one spot and a few more at another, without the slightest attempt at the usual order and system. Moreover, some cells contain single eggs, but others two, or even three, apiece. It looks as if some demented mother-bee from another hive had caught her keepers napping, and had made surreptitious excursion into the queenless stock. But the most careful search through the hive will reveal no queen, nor is one to be found. The explanation of the vagary is that one of the workers has, in some extraordinary way, succeeded in rousing her atrophied nature, and has become capable of laying eggs. Yet the doom of the colony is not delayed by this, but rather hastened; for these eggs will produce only drones, and thus still more useless mouths to feed. In one well-authenticated case, the bees of a queenless colony built a queen-cell, and actually transplanted to it one of these eggs laid by a fertile worker, a dead drone being afterwards found in the cell.

How the laying worker is produced under the spur of the national crisis can only be a matter for speculation, but probably the youngest bee of the colony is plied with the special food usually given to queens, and thus her generative faculties are, to a certain extent, developed.

CHAPTER X

A ROMANCE OF ANATOMY

THE modern commercial bee-keeper—the man who keeps his bees in hives of the most approved construction, all alike in colour and shape, and all in straight rows—is too prone to look only on the practical side of his work, and to regard with a certain ill-concealed contempt anything that does not directly promote what is, in his view, the one and only object of apiculture, that of honey-getting.

But with the bee-keeper who is also a bee-lover, the tendency is all the other way. To live in the very spirit of wonder, as he must who has once dipped down below the surface of hive-life, is to saddle but a slow, ambling jade for the race in material prosperity. In a bee-garden the habit of rumination comes on one like creeping paralysis, gradually but irresistibly. It is one thing, on a fine June morning, to start away from the house, pipe in mouth and busily trundling the honey-barrow, intent on a long day's work among the hives; it is quite another thing to keep industriously to the task hour after hour, when the sun has fixed his slothful golden grip upon you, and the drowsy song of the bees has worked its will on heart and mind.

Good resolutions have a way of petering out, reasonably enough, under these inviting circum stances. The honey-barrow makes the most comfortable seat in the world, and can be pulled up just where the shade of the linden-trees is thickest. Moreover, the blue smoke of tobacco, drifting lazily up through the sunshine, adds just that touch of deliberation needed in a scene where all is unmitigated, almost desperate toil; while what difference can it make if one alone be idle in the hundred thousand? And so, as often as not, the creaking wheel comes permanently to rest under the lindens; the honey is left to the honey-makers; the thoughts follow the bees into their hives, or may-be wend away over seas to the great plantations, where the dry weed filling the pipe-bowl was once a green leaf in an ocean of green, flecked over with blossom, and sung over by bees, whose ancestors might have come from this very nook in old England, where it is now all ending in smoke and quiet thought.

But, especially on rainy days, when there is much to do indoors—preparing the section-racks, discharging the honey from the full combs that, empty, they may be returned to the hives for refilling on the morrow, and what not—the tendency to set aside obvious, humdrum duties in beemanship has a still more capable ally.

The beeman with a microscope has given the seven-leagued boots to his conscience; he will never catch up with it again in a whole life's march. If the daily work in the hive, as seen with the naked eye, is a fascinating, duty-dispersing study, a microscopic acquaintance with the hive-worker herself, and the details of her extraordinary equipment, lets one into a whole new world of fact and thought.

It is only under a strong glass that the true place of the honey-bee in the scale of creation can be entirely estimated. Her work is evident to the most casual eye, but of the worker herself we get only a vague idea of a dim-hued, crystal-winged atom running a perpetual race with the wind and sunshine, or forming an all but undistinguishable speck in the seething, heaving multitudes within the hive.

But here, on the stage of the microscope, the honey-bee is revealed as a totally new creature; and, by little and little, a story unfolds itself about her which, in its way, is a perfect epic of life. No one can study the perplexities of hive-life for long without a conviction that a creature executing such varied and elaborate works must, of necessity, be herself highly developed in body and mind. But it seldom happens, even with the veriest tiro, that the expectation comes anywhere near the reality in such an examination of the common worker-bee. The unaided eye sees a creature, fashioned simply enough to all appearances—a brown, attenuated body, two pairs of wings, the usual six legs common to all insects, and a couple of bent horns, like threshels, that continuously waver to and fro. But under the glass this simplicity at once vanishes. From the tip of her antennæ to the barbed end of her sting, there is nothing about the honey-bee that is not made on the most bewilderingly, complicated plan.

Watching a hive at work on a busy day in summer, the attention is first drawn to the pollen-gatherers, labouring in by the thousand with the big, oval, brightly-coloured masses fixed to their hindmost legs; and it is first to the pollen-carrying organism that the glass is now naturally directed. The six legs, which looked all very much alike to the naked eye, are seen to be in three pairs, and the construction of each pair differs very markedly from that of its fellows. So far from their being simple legs, each has no fewer than nine jointed parts, and nearly every part carries a special piece of mechanism necessary and vital in the daily work of the bee. Whole treatises might be written on the functions of the human hand, yet the hand is a very simple contrivance compared with the legs of the honey-bee. The pollen-carrying device is on the thigh of the hind leg. The thigh is broadened out and hollowed, and round this oblong cavity is a fringe of incurving bristles which look as if they would hold anything. But before the pollen can be packed in these baskets it must be collected and kneaded together. Practically the whole body of the bee is used in pollen-gathering. Under the low power of the microscope it is seen that hardly any part of the trunk or limb is without its dense covering of hairs; but with the high objective these hairs cease to be hairs, and are changed into actual feathers, delicate herring-bone implements, which sweep up the pollen as the bee dives into the flower-cup for the nectar that lies below.

Nearly every joint of each leg is furnished with a comb of bristles, with which this pollen-dust is scraped off and transferred to the carrying-basket after being moistened by the tongue; while the hind-legs have each a complete, perfectly-fashioned curry-comb. Here the leg is widened and flattened, and covered on one side with nine or ten rows of short, strong spines, with which the bee scrapes her body just as a groom curry-combs a horse. At ordinary times she will carefully pack her load of pollen into its proper receptacles before returning to the hive, so that it shall be all ready for transference to the cells. At the cell-mouth she pushes each lump off by means of her other legs, leaving it to be rammed down into the cell by the store-keepers. No distinction is made here, every kind and colour of pollen being indiscriminately stored in the same cell; and when the cell is full, a thin layer of honey is smeared over all, to preserve it from the air. When, however, time presses, the bee will not stop to knead up the load, but will carry it home as it is, arriving in the hive smothered completely from head to foot as with gold-dust. Then the house-bees gather round her, soon scraping her free of her encumbrance, and she starts off again for another load.

The fact that insects can walk on both upper and under surfaces apparently with equal ease, is none the less remarkable because we see it going on every day of our lives. Yet the fly, crawling up the window-glass, or running about on the ceiling, owes his power of topsy-turvy perambulation to a very ingenious device. This is well illustrated in the foot of a bee. She has a pair of short, strong double claws, which will take her securely over all but the smoothest and shiniest surfaces; and it is with these claws that bees form themselves into dense clusters and knots and cables within the hive, holding hand-to-hand, as it were, in all directions. But when there is nothing for the claw to hold by, another part of the foot comes into play. This is a soft, flexible pad, which is always covered by a thick, oily exudation. In walking, the bee puts her feet down three at a time, the pads adhering instantly they come into contact with the smooth surface. At the next step the other three pads come into play, while the first three are stripped off. But each foot is capable of attaching and detaching itself independently of its fellows. In this case the stripping is accomplished by downward pressure of the claws of the same foot.

On each of her fore-legs the bee has an appliance which fulfils a very important office. It is a semicircular notch with a fringe of strong hairs, and when the leg is bent up, this notch engages with a curious projection on the next upper joint, forming an eyelet roughly circular in shape. With this exact and special tool she cleans her antennæ, and this is done at short intervals throughout the whole active time of her life, much as, in the operation of winking, the human eye is kept cleansed. The tongue also is freed from adhering grains of pollen by this device.

The question, How does a bee gather the flower-juices to make her honey? is met by certain popular naturalists with the assurance that she sucks them through a tube. This is so easy a generalisation that it amounts very nearly to positive error. The tongue of the bee is not a tube, as the word is usually understood. And she laps up the nectar as often as she sucks it. It depends entirely on the quantity to be dealt with; and a little careful dissection of the mouth-parts of the bee, by means of the microscope and a pair of long needles, will soon make the whole matter clear.

She is no beauty—the honey-bee, seen at such close quarters; unending toil, and a perverted, baffled nature, do not tend to loveliness in any of her sex. But her positive and almost terrifying ugliness, when looked at so disadvantageously, is soon forgotten as one comes to realise her abounding possession of that other kind of beauty—the beauty of utility.

To the naked eye her tongue is a bright brown, shining piece, protruding from her mouth, and hanging down with much the same appearance as an elephant's trunk. Under the microscope it is soon seen that this is not a tongue in the proper sense, but a continuation of the under-lip. It consists of six or seven different parts capable of being fitted together lengthways. There is a central part, longer than the rest, with a hairy spatula at its end, and when the other parts are closed about this, the whole virtually forms a tube within a tube. The spatula does the lapping when only minute quantities of fluid have to be taken up, and these pass into the mouth more by capillary attraction than by actual sucking; but when there is a brimming cup of nectar to be emptied, the whole mechanism of the tongue is brought into play. The longitudinal strips are placed together edge to edge, and the liquid is drawn out of the flower-cup by the action of the tongue-muscles in much the same way as water is lifted by a pump.

Now that we have the head of the bee under observation, many curious things about it can be ascertained. The strong, curved jaws, working sideways, are doubly interesting as the main implements used in the preparation of the wax, and largely in the comb-building. But the eyes and the long, flail-like antennæ rivet attention first. Whether the bee was made for her life, or the life—imposed on her by inexorable conditions—made the bee what she is to-day, the extraordinary adaptation of her physique to her environment is beyond all question. The great compound eyes, with their thousands of facets each pointing in a slightly different direction, are obviously made for wide and distant outlooks. It is with these eyes that the bee finds her way out and home over miles of country. In the worker the compound eyes occupy the whole sides of the head, but in the drone they are much larger, and meet entirely at the top. Thus, dallying in the sunshine, he is able the while to keep the whole arc of the sky under scrutiny, ready at an instant's notice to take up the love-challenge of the young queens.

But these large multiple eyes of the bee are of little use to her at close quarters, or in the deep twilight of the hive. For indoor use, and for near vision, she has three other eyes, containing a single lens each, and set in her forehead just above her antennæ. The popular belief, that the honey-bee carries on her busy life, and elaborate enterprises in complete darkness, is mainly a fallacy. Probably there is always some light, even in the remotest recesses of the hive—enough, at least, for the eyes of the bee, if not for our own vision.

The bee, however, would seem to depend very little on sight alone in the prosecution of her various tasks. There is little doubt that she possesses all the other four senses in a marked degree. Both the tongue and the lips have certain highly developed structures upon them which can be nothing else than organs of taste; while the most superficial acquaintance with the life of the hive must convince anyone that the bee possesses the senses of smell and hearing, and that very acutely. Where the seat of these two faculties lies is at present doubtful, and the exact functions of the antennæ are still a matter of conjecture. But it is at least certain that these latter perform vital office in every act or enterprise of the bee. It is obvious that the antennæ are very delicate organs of touch, but it is equally obvious that they are much more than this. It has been ascertained that they carry no less than six totally different kinds of instruments, each of which must have its distinct use.

Observation of the ways of the honey-bee has been carried on for thousands of years. More books have been written about the bee than perhaps of all other creatures put together. And yet our knowledge of her powers and organisation must still be reckoned in its infancy. The microscopists have dissected her antennæ and isolated all their various parts, but of the particular functions of these little or nothing is known at present. There are certain hairs, evenly distributed over the whole surface, which are presumably instruments of touch. But there are other hairs, or fine cones, which are hollow, enclosing a delicate nerve-fibre; hairs set loosely in a cavity; hairs curved and ringed, and of different lengths. Then there are mysterious pits and depressions, either open or covered with incredibly thin membranes, enshrining nerve-ends only just visible with the highest objectives. And the whole is linked up in an intricate nervous system that baffles every art and patience of research; while, when all has been investigated and described, no one is really any the wiser.

The antennæ are certainly touch-organs, and, in all likelihood, it is by their means that the bee hears and smells. Yet this only exhausts a few of their manifest possibilities. It is quite clear that we must admit the honey-bee to possess other senses than the five we know of; and—for a guess—some of these mysterious implements on her antennæ may be thought-transmitters and -receivers on the wireless plan. The wonderful unanimity of action among bees may be due to the fact that they can exchange ideas through the air, as men have now at last come to do. The faculty of speech, hitherto held up as man's insignia of lordship over the rest of creation, may be indeed a crude, archaic thing, compared with the mind-language of the honey-bees.

There is another conceivable function which the antennæ of bees may perform—that of unerring and instant estimation of short distances. They may be delicate measuring instruments, not mechanically applied in the way of a foot-rule or metric scale, but registering dimensions inherently, as our ears record intensity of sound. This would go far to explain how honeycomb is built, how the cells are made all of the same shape and size, although hundreds of the mason-bees are at work on the structure, not only at the same moment, but in succession, each bee coming and going in the murmurous gloom of the hive, and beginning instantly and unhesitatingly at the point where her predecessor broke off. As the central division of the comb grew, expanding in all directions downward, and the cells were built out horizontally at the same time, the bee would know by her sense of dimension when the limit of each side in the hexagonal cell-base was reached, and would know the proper angle to turn off at in the laying of the next foundation-line.

Anyone who has watched the flight of the bee must have been struck by its sheer facility and freedom no less than by its speed. It is quite evident that the bee is not only an accomplished aërial navigator, but that she sustains and propels herself through the air with very little effort. Obviously her equipment for flight must be a thoroughly efficient one, and yet at first glance it is not quite clear how she manages so well. The student of the flight-problem, taking his ideas and conception of first principles from the flight of birds, is accustomed to believe that there are at least two vital indispensable elements in the process—a pair of wings or combination of aëroplane and propellers that will sustain as well as drive, and some sort of steering-apparatus like the bird's tail. Yet, as far as a first general inspection carries us, the bee appears to have no rudder-mechanism at all, but to depend on her four wings for every purpose. The wings of the bird have a variable action. They can be used together or separately, and are as capable of eccentric adjustment, both in themselves and in relation to one another, as a pair of human arms. But the bee's wings have none of this adaptability. They have but the one motion, up and down; and they work symmetrically, each wing keeping time with its fellow. Yet the bee steers herself perfectly well in a hundred different evolutions, accomplishing all that the bird attains with his more complicated apparatus for flight.

The whole problem is bound up with another problem; and the two, difficult of solution apart, easily resolve one another when taken in conjunction. Insects are so called because their bodies are in two parts, entirely divided except for an extremely slender connecting-joint. We are so accustomed to accept this arrangement as a common fact in nature that we seldom stop to consider its real significance. It is not easy to see how such a construction can be anything else than a drawback to any living creature. But in the hive-bee the whole arrangement seems to amount to what must be called an ideal inconvenience, seeing that her honey-sac and complicated organs for producing the larval food are in her abdomen, with no way to them but through this fine joint. Clearly there is some weighty reason for it, outbalancing all other considerations, or it would not exist; and when we come to study it in connection with the honey-bee's peculiar system of flight, we soon arrive at the true solution.

It has been said that the wings of the bee have a perfectly symmetrical action, and that they have a single fixed direction, moving up and down, always at right-angles with the line of the thorax. Under the microscope each of the four wings is seen as a transparent, impervious membrane, intersected with fine ribs. The front wing, however, has a much stronger and stiffer rib running the entire length of its upper edge, and it is on this main rib that almost the entire force of the flight-muscles is concentrated. If you look farther, you will see that the under wing has a row of fine hooks along its top edge, while the lower edge of the upper wing is flanged or folded back. In flight the hooks on one wing engage with the flange on the other, and thus the wings on each side are automatically locked together, forming one continuous air-resisting surface. This combined wing is very flexible throughout, except at its upper edge, where it is stiffened by the main rib. In action, therefore,—the force being applied practically to the edge alone, which resists the air while the rest of the wing bends to it—the result is that the whole wing becomes an oscillating, inclined plane, whose inclination, forward on the down-stroke, is still forward on the up-stroke, because the plane-inclination reverses itself automatically.

From this it will be understood how the flexible wings of the bee are used in straightforward flight; but, seeing that the wings themselves are incapable of independent or irregular action, it is not yet clear how the bee contrives to steer herself, rising or descending, or turning sideways, just as the mood seizes her. It is here that the reason for the peculiar construction of her body becomes plain. The fine link which unites her abdomen to her thorax is really an universal joint, actuated by a series of powerful cross-muscles, and the bee steers herself through the air by using the weight of the lower half of her body as a counterpoise. By swinging her heavy abdomen forward or backward, or from side to side, she changes her centre of gravity, and the line of force of her aëroplanes, at one and the same time. Actually her body keeps its vertical position, being her heaviest part, and it is the lighter wing-supporting thorax which is deflected. But the result is the same, and every variety and direction of flight is accomplished by the bee on what seems a far more simple plan than that evidenced in the flight of birds.

One of the most difficult things to account for in the life of the honey-bee is the fact that the temperature of the hive can be varied at the will of its occupants. The system of mechanical ventilation will, of course, explain how the hive is kept cool in the greatest heats of summer, but it does not explain the sudden accessions of heat to which it is liable from time to time. These occur principally when the wax is being generated. Under the bronze armour-plates of her body the worker-bee has six shallow, but broad depressions, beneath which the wax-glands are placed. Perfect rest and a high temperature seem to be necessary for the stimulation of these glands, and the wax-makers consume a large quantity of sweet-food during the process. It is generally stated that bees fill themselves from the stores of mature honey before uniting in the cluster; but it is more probable that the food consumed during wax-making is principally the nectar, almost as gathered from the flowers. This view is confirmed by certain experiments which were undertaken to decide the amount of food assimilated during the production of a given weight of wax. When the bees had access only to honey, it was found that five or six pounds were needed during the time that one pound of wax was produced. But if the bees were fed on a plain syrup of cane-sugar, more wax was generated. The chemical composition of fresh nectar is almost identical with that of sugar from the sugar-cane, but mature honey contains practically no cane-sugar at all. It is very doubtful, therefore, if the economic bee would deplete her hard-won stores of honey for a purpose that could be better accomplished in another and cheaper way. And it should also be borne in mind that the natural time for comb-building coincides with the season when nectar is in greatest plenty.

These sudden variations in temperature appear to be brought about by a wholesale increase in the rate of respiration among the bees; and there is nothing that excites the wonder of the student of hive-life more than the breathing-apparatus of the bee, as seen under the microscope. Practically her whole physical system is directly supplied with air, drawn in through her many spiracles. As far as scientists have been able to determine, there is not a fibre or nerve in her entire body that is not reached by the minute ramifications of the air-ducts, in direct communication with the great main breathing-vessels in the bee's abdomen. Respiration appears to be largely voluntary with the honey-bee. She breathes only when the necessity for it arises, and will sometimes arrest the action entirely for three or four minutes together. But when the wax-making is going forward, or swarming-time is near at hand, the quick, vibratory movement of respiration is visible everywhere in the throng of bees, and the temperature of the hive climbs up often to a dozen degrees above its normal point.

The breathing system of the honey-bee is closely connected with her sound-organs. Anyone asked to describe the note made by a bee would probably say that she hums or buzzes, and there would be an end to most ideas on the matter. But to the beeman this is a pitifully inadequate statement of the truth. The bee comprises in herself not one, but a whole choir of voices, and she has a compass of at least an octave and a half. Every one of her fourteen spiracles, and each of her wings, is capable of producing sound; and these sounds can be endlessly varied in quality, intensity, and pitch. It is no exaggeration to say that the honey-bee is as accomplished a musician as any bird; but as each individual voice is for the most part lost in the general symphony of the hive, it is difficult to get a complete idea of her capabilities as a soloist.

The voice-apparatus in the spiracles is one of the most intricate things in the whole anatomy of the bee. It has a multiplicity of parts, and is obviously designed to convey a great variety of sounds. The wings also produce tones that run up or down in the scale, according to their rate of oscillation; and from them comes the sibilant note usually called buzzing. Listening to the hive-music at any season of the year, it is impossible to resist the thought that bees not only hold individual communication by means of these infinitely varied sounds, but that the general note given out by the multitude unerringly expresses the state of affairs within the hive for the time being. A prosperous stock voices its busy contentment in a way impossible to misunderstand. It is a deep, blithe, resonant sound, like the steady running of well-oiled machinery, each wheel adding its own whirring melody to the general theme. Weak or famishing colonies give out a wavering, intermittent note, the very voice of complaint and fear for the future. When a hive has lost its queen, a capable bee-master should have no difficulty in divining the trouble by listening at the hive-entrance. A queenless stock is all clamour and the hubbub of divided counsels. The ordinary rich reverberation of labour stops, and a sound of panic goes to and fro in the hive unceasingly. If a hive be quietly opened, and its queen removed with little disturbance, it may be some time before the bees discover their loss. Some colonies experimented with in this way realise their deprivation immediately, and the hue-and-cry begins at once. But one of the most curious facts in bee-life is the variation in intelligence, and alertness of perception, between the different hives. A steady-going, dull race may be a considerable time before it perceives the absence of its queen. The common note of work goes on unchanged until the fact dawns on it. And then the peculiar shrill outcry commences, overpowering all other sounds until reason again asserts itself in the colony, and the bees set about the work of raising another queen.

The voice of the drone is deeper and hoarser than that of the worker-bee, by reason of his larger body; and his noisier buzzing is explained by his greater length and breadth of wing. The queen also has a deeper, more husky voice during flight; but she has, in addition, a peculiar cry of her own, an old familiar sound to bee-keepers all the world over. It is heard principally just before the swarming of the hive. Certain old skeppists profess to be able to foretell the date on which a swarm will issue by studying the cry of the queen. On quiet nights, just before the swarming-season commences, it may frequently be heard above the general murmur of the hive by bending the ear down to the entrance. It is a shrill piping sound, repeated over and over again, and often answered by other and fainter notes. How it is produced is not certainly known, but probably it is caused by the wings or legs being sharply rubbed together, much as a cricket or grasshopper utters its cry. The louder note is made by the old queen, and there is no doubt of its import. Jealousy and the lust of battle are on her, and she is trying to get at the young princesses in their cells. The cry is one of baffled fury as she strives with the guards about the cells, and the answering notes come from the imprisoned queens who are just as eager for the fray. The old skeppists are never far out in their reckoning. When this state of affairs has begun, the crisis is imminent; and the morrow is sure to see the emigrating party setting off for its new home, carrying the old queen irresistibly with it.

It has been said that the nurse-bees, who have the entire charge and care of the young brood, feed the larvæ from their mouths with a thick white fluid, which is aptly called bee-milk. All the time the nurses are engaged on this work, they are themselves hearty eaters of both honey and pollen; so that at first sight it appears as if the bee had the power of instantaneous digestion, feeding herself at one moment, and, at the next, regurgitating this food, changed into a totally different substance, to feed the young grubs. Moreover, there is another wonderful thing regarding this bee-milk. It has been proved by careful analysis that its composition varies considerably. The male, female and queen-larvæ are all fed with it, but its constitution differs, not only with each kind of larva, but according to the age the larva has reached. The bee must therefore have her whole system of digestion under full voluntary control. How she manages this critical part of her work can only be understood by the aid of a good microscope.

The Bee-Nursery: Tending the Young Brood

Perhaps there is nothing more wonderful, in the whole wonderful anatomy of the bee, than her digestive organism and its contributory system of glands, each of which has its special and important use. When she draws up the nectar from the flowers, it passes at once into the first of her two stomachs, which is simply and solely a reservoir. Here it can remain indefinitely at the will of the bee; or it can be thrown up and poured into the comb-cells, to be brewed into honey; or it can be allowed to pass through a valve at the base of the reservoir into the bee's second and lower stomach, where digestion takes place and the honey and pollen are formed into chyle. But, by one of the most ingenious devices in nature, this second stomach is also capable of returning its contents to the mouth, and the chyle is there changed into bee-milk for the nourishment of the larvæ.

The worker-bee has, in all, four distinct glands, each secreting a fluid with properties different from the other three. These glands are all situated in the mouth. Two of them have a common opening in the upper side of the root of the tongue; and as the bee sucks, their combined secretions mingle with the flower-juices automatically, and the first step in the change of the nectar into honey takes place. The third gland is in the roof of the mouth, and it is the secretion from this gland which acts on the regurgitated chyle, and changes it into brood-food. The fourth gland is double. These twin-glands have their openings at the base of the jaws, and the action of chewing is necessary to excite their secretion.

The valve between the upper, or honey-stomach, and the lower, or chyle-stomach, has an extensible neck, and the bee can, at will, raise this telescopic piece through the interior of the honey-sac until the valve is pressed against the opening into the gullet. Thus the contents of the lower stomach can be driven into the mouth without coming into contact with the stored sweets in the reservoir, and this pre-digested matter is always ready at an instant's notice for the use of the larvæ, or for the nourishment of drones or queen.

It has been said that the nursery-work of the hive is undertaken exclusively by the young bees during the first fortnight or so of their lives. After this time they make their first foraging expedition, beginning with pollen-gathering, and relinquishing this in turn for the collection of nectar when they have arrived at full maturity. The mature workers take no part in the feeding of the larvæ, except on very rare emergencies. In relation to this, it is a curious fact that the gland in the roof of the mouth, which acts on the chyle, forming it into brood-food, is in full development only during the first weeks of the worker-bee's career. After that its activity swiftly declines, until, in old workers, it becomes largely atrophied.

The digestive gland-system of the honey-bee, although it has been fairly well explored by the scientific naturalists, is still much of a mystery, and this especially with regard to the glands attached to the jaws. The secretion from these glands—obviously a very powerful acid—is mainly used to convert the raw wax from its hard, brittle character into the soft, ductile material of which the combs are made. It is probably used to some extent, also, in the preparation of the brood-food, in conjunction with the gland in the roof of the mouth. It mingles with the pollen when this is masticated, and no doubt it has various other uses; but no one seems as yet to have discovered why these two glands should be so enormously developed in the queen, who takes no part in the nursery-work or comb-building. The whole question will naturally have little more than a passing interest for the general reader; but, to the bee-keeper with a microscope, it takes a prominent place among the debatable things in hive-life. If the difference between the queen-bee and the worker-bee—a difference of organic structure as well as mere development—is really brought about by variation in the quality and quantity of the food supplied to the larvæ, then the action of these glands cannot be over-estimated in importance, and cannot be studied too deeply: they form the very spring and fount of life. Yet is it certain that the influence brought to bear on the young grubs by the nurse-bees is wholly restricted to the matter of food? The worker-bee has several curious organs and gland-systems in various parts of her body, in addition to those already enumerated, to which no rational use has yet been assigned. The more we study her extraordinary equipment, the less justification there appears to be for dogmatising about her, limiting or particularising the function of any one gland or implement in the whole unending array. The old adage, that there is nothing invariable about the honey-bee, is like to be as true with regard to her physiology as it is with her habits of life; and, for all we can tell, to-morrow's knowledge may render obsolete much of the carefully garnered knowledge of to-day.

If the story of the honey-bee's anatomy has everywhere some of the elements of romance about it—in its unexpected incidents, its adventurous colour, its shadow of a great design—this spirit suffers no abatement when we come, in a last view of it, to consider her as one carrying arms, one bearing such a weapon of offence as never came into human mind to fashion. The long curved scimitar of the queen, which she cherishes so carefully that nothing will induce her to strike with it except when it is to be turned against a royal foe, is otherwise little else than a harmless piece of domestic furniture. But the sting of the valorous worker-bee, seen under a microscope, is a positively terrifying engine of destruction. Popular science generally describes it as a sheath containing a barbed and poisonous dart; and the trite comparison is always made of the bee's sting with the finest sewing-needle, the latter being likened to a rough bar of iron. The idea of a sheath is pure fiction, as a little painstaking examination will soon reveal.

The bee's sting is made up of three separate lances, each with a barbed edge, and each capable of being thrust forward independently of the others. The central and broader lance has a hollow face, furnished at each side with a rail, or beading, which runs its whole length. On the back of each of the other two lances there is a longitudinal groove, and into these grooves fit the raised headings of the central lancet. Thus the sting is like a sword with three blades—united, but sliding upon one another—the barbed points of which continue to advance alternately into the wound, going ever deeper and deeper of their own malice aforethought after the initial thrust is made. It is a device of war, compared to which the explosive bullet is but a clumsy brutality. Yet this is not all. To make its death-dealing powers doubly sure, this thorough-minded amazon must fill the haft of her triple blade with a subtle poison, and so contrive its sliding mechanism that the same impulse, which drives the points successively forward, drenches the whole weapon with a fatal juice.

The tendency to be unduly scientific, to meet these things with exact and unimaginative interest, receives its final quietus here. For he who realises the whole deadly efficacy of the honey-bee's sting cannot logically pass it by as a mere remarkable provision of nature, praising God for it complacently, but must concede it a much wider significance. This complicated weapon of the stunted, sex-perverted worker-bee owes its existence as much to deliberate art as to nature, or those who watch the Omnipotent in hive-life are strangely and perversely led astray. In the queen-mother, whose physical organism may be said to be comparatively unchanged from its aboriginal type, we see the part corresponding to the worker's sting, essentially another creation. The queen's ovipositor is longer; it is curved; the barbs upon it are small and insignificant; the fluid in the secreting-gland is no poison at all, but a thick opaque substance, whose true use is probably to glue the eggs safely to the bottoms of the cells. She is also provided with a pair of blunt instruments covered with sensitive hairs, which serve, with the ovipositor, to guide the egg securely to its destination. The worker-bee has these feelers on either side of her sting, but she has perverted them to a very different office, that of seeking out the vulnerable parts of her enemy. And what a drastic change her will, or that of her foster-mothers, has wrought in the whole contrivance! She has bartered the privilege of motherhood and years of life for a few short months and a share in the communal sovereignty. She must be ready to further the well-being of the hive by the art of war as well as by the arts of peace. Therefore she has deliberately helped in fashioning the ploughshares into cannon. A little change in her food as a nursling, an infinitesimal leaking from a gland that takes the full power of the strongest glass to see,—and, with all the other multitudinous changes of form and character, this last miracle comes quietly into being. The egg-depositing shaft grows short and straight; its moderate indentations become cruel jagged barbs designed to hold as well as to kill; the harmless, egg-fastening gluten is quickened into a virulent poison; and the death-dealing thing is ready and ripe for service against all honey-lovers, the hereditary foes of the hive.

CHAPTER XI

THE MYSTERY OF THE SWARM

THE old "swarm in May," beloved of ancient bee-men, is rapidly becoming a thing of the past. Modern hives and modern methods, although they have not as yet achieved their main intent of abolishing natural swarming altogether, yet tend to bring this extraordinary ebullition of hive-life to its fulfilment later and later in each year. Far from being a virtue, as of old, an early swarm, or indeed any swarm at all, is now accounted a misfortune, even a downright disgrace, in scientific beemanship. And yet the bees, though easy to discourage, are hard to teach. In spite of roomy hives and a watchful bee-master ready to give them an unbroken succession of young and fertile queens, and a whole houseful of new furniture at a moment's notice, still the bees go on playing this mad game of wholesale truantry, and still the bee-keeper must stand looking hopelessly on from the midst of his elaborate appliances, while his property sings about his ears, or wings away into the upper skies, irrevocable as last year's mill-water.

Bee-Men call it the swarming fever; and fever it is in very truth. The reasons for it have long ago been crystallised into exact and accepted phrases. An overcrowded condition of the hive; the desire of the bees to get rid of a failing queen; the excitement of the queen herself at the menace of coming rivals; the natural instinct of colonies to increase and multiply—anything but the one all-sufficient and obvious reason, that bees swarm because they suddenly and intensely desire it.

A Swarm in May

The story of the Sioux Indian,—won for civilisation from boyhood, over-educated and over-refined, decorated with a high college-degree and adorning a great pulpit, and then casting it all to the four winds, stripping and painting himself, and raging away with his kind on the war-trail,—has a near parallel in the behaviour of bees at swarming-time. Instinct could never be a party to such an inconsequent, outrageous, brilliantly reckless, joyous proceeding. But it is ever in the way of reason to be splendidly unreasonable at times, and here the honey-bee shows herself the true child of her origins. From a stern, self-elected destiny-maker, callously pressing to the forefront of life over all obstacles of heart and hearth, she changes back, for the nonce, into the aboriginal bee-woman, thoughtless, pleasure-loving, improvident, spending the garnered treasure of laborious days in the one mad moment's frolic.

For it is impossible to regard the incident of the swarm as only one more link in the chain of sober, calculating bee-wisdom. It is obviously a lapse, a general falling away from the all-wise, public polity. For a single hour in her drudging, joyless, perfect life, the worker-bee battens down all the virtues, and rages forth like the Sioux Indian to swill at the stream of forbidden love and laughter, unmindful of the cost. Just when the common self-abnegation is yielding its rich first-fruits of prosperity, and the hive is overflowing with its wealth of citizens and possessions, this fever comes among them, and spreads like a prairie fire. By all laws of prudence it is now, of all times, that every child of the Mother-State should stand by her mightily, to uphold her in the high place won for her by unending toil and innumerable lives. But old ancestral memory wakens, calling irresistibly. Nature, in the beginning of time, made the honey-bee to inhabit a tropic land, where there was no need for pent, cold-withstanding houses, nor any use in laying up provender for days of dearth, because the land flowed with perpetual honey. Bee-life in those far-off ages was all dancing in the sunshine, and the bee-woman had little to do but to fly to the nearest brimming flower-cup when her nurslings wanted food. But a cooling world, the ever northward trend of her race, and then the folly of her own wisdom—intellect turning upon itself—all combined to lose for her the old slothful paradise of plenty. The drone, reasoning inversely by the wisdom of his folly, made a better compromise with fate. He held to his life of ease and his gratuitous pleasures at all cost, and let his mate go her way undeterred, blinding his eyes to the new necessities. Work and responsibility gradually soured and sharpened and hardened the one, while dependence on his womenkind as insidiously changed the other into a creature of idleness and the senses. And when he came at last to realise the outcome of it all, it was too late. The matriarchal commonwealth was established, hedged round securely with a myriad poisoned blades. To live a drone had been his heart's desire, and now dronehood, mere seminality, was allotted to him as a retribution. The things for which man lifts his unregarded prayer all his life through, might very well prove his fittest punishment, granted to him in the Hereafter: so little can man or drone distinguish between the enduring things of life and death.

But of all intolerable fates, that must be least bearable, to have wisely willed and beautifully fashioned our own eternity; and then, being only human, or at least reasonable, to find its goodness really smooth-going, colour-fast, impregnable at all points, with never a bright break or flaw to vary the monotony of well-doing. No wonder the honey-bee swarms, breaks helter-skelter out of her prison-bounds of order, commendable toil, chill, maidenly propriety; and goes rioting away for one short hour of joyousness and madcap frolic, such as her primæval sisters looked to as the common day's lot, when there were no hives, and motherhood was not the sole prerogative of one in thirty thousand, and when the sun burned high and cheerily in heaven from end to end of the tropic year. It is easy to be wise, and temperately scientific, in accounting for this feverish impulse of the worker-bees, allotting it a sound and circumspect part in the furtherance of the general polity. But is it not, in the main, Nature—the atrophied sexual spirit—awakening, or at least stirring a little in her age-long sleep? In the sultry August evenings the young queens of the ant-hills pour out in unnumbered thousands to meet the males, and people the ruddy sunshine with the glint of their wings.

This is swarming in its truest sense. The wingless, workful, underground existence follows, but the love-flight of the ants, while it lasts, is none the less a real, intensely joyous thing. And surely the swarming-fever that so strangely and inopportunely seizes upon hive-life, is at one with it in nature and spirit, although its original purpose and value have been long ago lost in the ages.

The one in the whole multitude who alone has the full inheritance of her sex, the queen-bee, seems often at the fountain-head of the revolution. Sometimes, undoubtedly, it is she who first develops this longing, feverish unrest, and by little and little communicates it to the whole colony. Here the variability of bee-nature comes sharply into evidence. Some hives will show this restless spirit for many days before the swarm issues, while with others the great upheaval seems, as far as the mass of bees is concerned, to be a sudden unpremeditated thing occurring in the midst of the universal content and industry. The preparations for raising new queens are always taken in hand betimes, but probably this is the work of the far-seeing, sober old bees of the hive, with whom communism has become a settled and accepted calamity. The bees who will ultimately constitute the swarm may be supposed to nourish their secret desires from the first moment the queen shows signs of mutability; to neglect all their old tasks, first in heart and then in reality; and finally—when the queen's mood has reached its culminating point, and her work in the hive is in virtual abeyance—to throw down plummet and trowel and hod, and rush forth in a wild, hilarious company, urged by a longing that they are as powerless to resist as to understand.

In the study of bee-life one comes upon many questions, but seldom answers to fit all. If the queen's fecundation takes place only once in her life, and nature intends this to suffice for her whole fruitful period, it is not easy to see why she should go out with the swarm at all. That she is not the inveterate recluse as generally believed, and that she does occasionally make short flights in the open during her laying career, is well proved.

A Mammoth Swarm

The desire, therefore, to see the light again after a long incarceration cannot be urged as her reason for going off with the swarm. A much more plausible notion is that the sexual spirit is again roused in the queen, just as it seems to be roused for the first time in the worker-bee; and that, with all, the journey is undertaken as a mating-flight, a faint re-echo of a racial custom long extinct, bearing the closest analogy to the marriage-swarm from the ant-hill. It must be borne in mind that, although the queen-bee is undoubtedly rendered capable of producing her kind of both sexes during several years, as the result of a single fertilisation, it cannot be incontestably held that she never again meets the drone under any circumstances. There is nothing in her physical organism to prevent a second coition, although with the drone this is impossible, for more reasons than the all-sufficient one—that he dies in his marriage-hour.

In the old bee-gardens, where the "swarm in May" is still a living, present thing, it is pleasant to sit with the proprietor under the rosy shade of apple-boughs waiting for the swarms to issue, and "talking bees," which is the most nerve-soothing, soul-refreshing occupation in the world. There never was a bee-keeper, new style or old style, too busy to talk, provided that you met him with understanding, and were as impatient as he of digressions from the all-important theme. One soon gets tired of imparting information as to the wonders of hive-life to the ignorant and plainly apprehensive stranger, and none sooner than he of the old school. In the quietest apiary of purebred English bees there are always a few individuals of crotchety nature, who will search you out in the shady orchard seat, and, as like as not, knife you on the least provocation. If you are a beeman, you treat these vindictive approaches with unconcern. You go on listening to the old man's talk, while the bee shrills away at your eyelids, or creeps into your ear and out again. If you keep quiet, she will soon relinquish the dull sport, and wing harmlessly away; and the thread of the master's discourse is not interrupted. But the uninformed stranger is a nuisance at these solitudes for two. He flinches and shudders; makes little irritating retreats; beats about wildly with his hands; or, if he is made of the sternest metal, he sits rigidly upright when he should be reclining at his ease, and turns such a painfully polite, though distracted, ear to his informant, that the stream of talk is sure to dry up incontinently, and he feels as little welcome as ghostly Banquo at the feast.

When you have once lived among hives it is a sore thing to be without their music. On warm days, winter and summer alike, there is always this drowsy, dreamy song in the air; and dancing without the fiddlers is no more depressing an occupation than, to a beeman, is loitering in a garden of mere silent vegetables and flowers. Sitting now under the bower of apple-blossoms and watching for the swarms, the full sweet note from the hives comes over to you like the very voice of serene content. It pervades the sunshine. It gently qualifies the slow wind in the tree-tops. It lifts and falls like the lilt of a far-off summer sea. This is the labour-song: the song of the swarm is very different. To the trained ear the cæsura that presently comes in the midst of the music is as clear as a pistol-shot, though you may detect no change. The old bee-keeper stops short in his wandering tale about famous honey-years of half a lifetime back, seizes key and pan, and hurries across the garden. It is the old green hive again, he tells you, as you press hard upon his heels—it is always the old green hive that has swarmed the earliest every May for years back. And forthwith the key and pan begin their clattering ding-dong melody.

Old-fashioned bee-keeping is not always a matter of straw. Box-hives, without, of course, the modern inside furniture, have been in use nearly as long as the straw skep; and the hives in the garden are of this ancient pattern. The old green hive is keeping well up to its reputation. Already it is the centre of a swirling crowd of bees, and, as you look, a dense black stream of them is pouring out of the entrance so fast and furiously that it is almost impossible to distinguish what they are. And the old wild trek-song is growing louder and deeper with every moment, a rich vibrant tenor note unlike any other sound in nature. There is no doubt at all of its import, as you stand in the wing-darkened sunshine, caught up in the excitement of it all, and feeling much as if you were facing a tearing sou'-west gale. Every bee of the twenty or thirty thousand volleying madly to and fro overhead, is singing her bravest and loudest. There is only one meaning to the whole gargantuan chorus. It is sheer jubilation melodised: a wild, glad song of freedom, as though not a bee amongst them had ever before set eyes on the sunshine and the wealth of an English May.

The great door-key, a ponderous, antiquated piece of metal, beats out its clanging note, and the swarm lifts higher and higher into the blue. Gradually the sombre mist of bees draws closer together, looking now like a little dark cloud strayed from a forgotten summer storm. Now it sails slowly northward, and lightens, as the sunlight is caught by the beating wings as in a net of silver; and now it veers away into the very eye of the sun, and changes into black, revolving tracery again; whirring wheels within wheels of insect-life, spinning-wheels making thread to weave the garments of a whole nation, and humming as never spinning-wheels hummed before.

Hiving The Swarm

But the beginning of the end is nigh; the time of singing is nearly over. The old beeman stops his weird tom-tomming, throws down key and pan, and points to the topmost branch of a young apple-sapling. You see a little black knot of bees clinging to it no larger than a pigeon's egg. A moment later, and it has grown to the size of a double fist, and another moment sees it twice this size again, as the flying bees stream towards it from all directions. Now it is as big as a quart measure, and the branch is slowly bending down under its weight. In an incredibly short space of time the whole swarm has joined the cluster; they hang together in a long, brown, glistening, cigar-shaped mass, well-nigh touching the ground, and the wild, merry music is over for good.

Gently swaying in the sunlight, lifeless and inert but for a few restless bees that hum about it, the sight of a settled swarm has an almost uncanny effect on most observers. A little before, the whole garden was filled with its deafening, joyous hubbub; now a strange silence has fallen, and it is impossible to dissociate from its present state the idea of an abject depression and disillusionment, as though the whole thing had been but a mad escapade, of which the bees were now heartily ashamed. If we may conceive the issue of a swarm to be a freak of ancestral memory, the sudden irresistible impulse to follow an old racial habit, long obsolete, it is not difficult to account for the obvious change of mind that has now come over the absconding host. Packed within the hive in a feverish, surging multitude, disabilities were not self-evident as they are now, tried in the light of day.

> "Violent delights have violent ends,
> And, in their triumph, die."

And now there is the morrow to be thought of: life to be rendered possible in all odds of weather; a home to be made; the queen-mother to be sheltered—she, the one remaining possession of the crowd, beggared now, but so rich a moment before. There is hard work ahead, enough to sober the giddiest among them. The madness has gone as quickly as it came, and now the honey-bee is to show herself a reasoning creature, if never before.

It is believed by most bee-keepers that a swarm selects the site of its future dwelling some time before the expedition starts, in many cases several days earlier. An old trick among cottagers is to place out empty hives in their gardens, and these not uncommonly attract errant swarms. A few bees are seen cruising about, and subjecting the hives to a close scrutiny. These pioneer bees disappear, and after a variable time, from a few minutes to a few hours, or even days, a whole army of bees suddenly descends from the sky and takes possession of the new home. When the interval between the appearance of the scouts and the arrival of the main body, is only a short one, the reconnoitring bees have been manifestly sent out by the clustered swarm; but in the case of long periods elapsing, the scouts must have been sent in search of the new location before the swarm issued. Probably, although the bulk of the party is imbued with this reckless spirit alone, thinking and caring for nothing else but the escape and the frolic, many of the older and wiser bees undertake the matter in a temperate, businesslike way, as they would go about any other important hive-operation. In one sense, therefore, the old notion of there being "subordinate lieutenants, captains, and governours" in a hive may not be so very far from the truth. That these scouts are actually sent out to find a suitable site for the new colony, either before the swarm leaves or while it is clustered in the open, is a well-established fact, so that some of the bees at least must keep their wits about them throughout the general chaos.

And with these wiser virgins must be reckoned the queen, in spite of the fact that she joins in the public excitement and restlessness. For some days before the great emigration her work of egg-laying is largely arrested, and this retentive action renders her so heavy and bulky that often she can scarcely get on the wing. The object of this is that she may be all the more ready for laying when the new home is established. It is also well ascertained that all swarming bees have their honey-sacs well filled, and this loading up for the journey takes place just before the signal for departure is given. There is great variation in the behaviour of the different stocks in a bee-garden during the swarming season, and many close observers are unable to detect any sure signs that a particular hive is going to swarm. But it appears fairly well established that, when a swarm is imminent, nearly all the bees of that stock remain at home, even when all other hives in the garden are in full foraging activity. Such a hive gives out a peculiar throbbing note, which suggests the noise made by a powerful locomotive brought to a standstill, but with full steam up, and impatient to be gone. Just before the issue of the swarm there is often a curious lull in this pent-up, forceful sound, and probably this is the moment when the travellers are lading themselves up for the march. Immediately after—and here it is difficult not to believe that a definite, authoritative signal for the movement is given—a sudden stir and tumult begins in the centre of the crowded hive, much like that caused by a heavy stone cast into water. This radiates swiftly in all directions until it reaches the bees near the entrance, and then the general rush for the daylight starts. Where a hive is much overcrowded there will already be a cluster of bees numbering many thousands packed tightly together on the alighting-board, and sometimes covering the whole face of the hive. But this mass melts away directly the swarming begins, the waiting bees taking wing all but simultaneously with the others.

It was anciently believed that the queen led the swarm, but this view is not borne out by modern observation. As often as not half the bees are on the wing before she makes her appearance, and sometimes she is among the very latest to leave, or she may decide at the last moment not to go at all. In this case the bees do not cluster, but after a few minutes' wild tarantelle in the sunshine they all troop back to the hive.

When once the swarming-party has gone off, the old hive seems to settle down to its ordinary occupations as though nothing out of the way had happened. The congested state of affairs no longer exists, but otherwise the work of the hive is proceeding in the usual way. The bees left behind are mainly young workers who have not yet commenced foraging, but there is always a fair sprinkling of old workers and drones. Generally the hive is queenless for the time being, the new queen not having yet broken from her cell. There may be four or five queen-cells in various stages of development, or rarely as many as a dozen. Sometimes, however, the first of the queens will be already hatched and wandering over the combs, meeting, as usual at this stage of her career, perfect indifference from all she encounters. But hives have been known to send off a swarm when the preparations for raising a new queen have been scarcely begun. So variable is the honey-bee in all her ways.

If the objects of swarming were merely to relieve the congestion in the hive, and to change the mother-bee, the whole thing should now be at an end. But the swarming impulse is rooted in far deeper soil than mere expediency. With some strains of bees the fever seems to die out after the one attack, and the stock settles down quietly to work for the rest of the season.

But more often than not this first taste of adventure serves only to whet the national appetite for more. About nine days after the first swarm leaves another swarm often follows, and this may be succeeded by a third or even a fourth at a few days' interval, resulting in some cases in the almost complete extinction of the stock. The old skeppists called the second swarm a "cast," the third was a "colt," and the fourth a "filly."

It is difficult to understand how, in a community where individual interest is so ruthlessly sacrificed to the general good, this self-destructive policy should be permitted. But taking the view that swarming is in the main a vague and incomplete resurrection of a long obsolete habit in bee-life, a workable theory at once suggests itself. Under primæval conditions the continued life of the mother-colony may have been unnecessary. Its purpose may have been fully served when a number of young queens and drones had been raised, and the whole had swarmed out together, each to form a new settlement. It must be remembered that the bee-hive, persisting indefinitely from year to year, is really quite a modern creation, and became practicable only with the invention of the movable comb-frame, which allowed the bee-master to effect the renewal of combs.

The Swarm Hived

It has been seen that the brood-combs get gradually choked up with the pupa-cocoons, which each bee leaves behind it. These webs are so incredibly thin that a dozen of them make little appreciable difference to the capacity of the cell, and combs have been known to remain in use for brood-raising as long as twenty years. But eventually they must become useless; and then, as bees do not, or cannot, remove old combs to make way for new, the community must leave for a new home, or gradually die out. Thus the age of the old hives was definitely limited.

Modern beemanship has wrought many other changes in the life of the honey-bee in addition to creating the permanent hive-city. The number of bees in a single strong stock, housed in a modern frame hive, is probably three times as great as that of a wild colony. The work of the bee-master affects almost every aspect of bee-life, enlarging the scale and the scope of all that the bees attempt. The result of this is seen not only in an increased population and more extensive works, but in a change in the very systems of life. Plans that work very well on a small scale do not always succeed on a large. The sanitary problems of a city are necessarily very different from those of a village, in principle as well as in degree. And probably much of the ingenuity of system and device observable in modern hive-life is directly due to human agency, the new conditions introduced by the bee-master serving to educate the bees to greater effort and resource.

The behaviour of these after-swarms offers a curious contrast to that of the first one. If it were possible to point to one fixed and invariable law in bee-life, it would be to the fact that a prime swarm will leave the hive only on a fine, warm day, and generally about noon. But casts and colts and fillies seem to take no count of time or weather, issuing just as the mood besets them, early or late, and caring nothing, apparently, for the conditions abroad. It is even on record that once a second swarm came off at midnight, when the moon was at the full and the weather very clear and warm.

There seems altogether much more method in the madness

that seizes on a colony swarming for the first time, and if thereafter the hive settles down to its old courses, the national character for sobriety and industry soon rehabilitates itself. But it is just the strength of this public inclination towards order and labour which varies so greatly in different hives. How matters are likely to go can be readily ascertained by setting careful watch on the hive from the day the first swarm leaves. There are sure to be several queen-cells, some capped over and almost ready to hatch out, and others in various stages of development. All these cells are constantly and assiduously guarded by the worker-bees, because directly one of the queens is hatched, her first thought is to make a speedy end to all future rivalry by murdering her sisters. She comes from her cell evidently spoiling for a fight, and imbued to the core with that inveterate hatred of her kind which is the ruling passion of her existence.

That worker-bees and queen-bees should have an identical origin, and yet that the nature of the one is to live in perfect harmony, while the nature of the other is to be at perpetual war, is one of those mysterious things in bee-life which probably will never be explained. If the queen-bee of to-day can be really taken as an approximate type of the aboriginal female of her race, it is not difficult to understand that after her generation in force the communal life of the mother-stock would become an impossibility, and that with the mating-swarm its natural existence was brought to a close, much as we see it happen in wasp-life.

It is during the quiet nights, after the issue of a swarm, that the peculiar shrill voice of the queen is most frequently heard. As she strives with the guards that surround the cells of the other young queens as yet unliberated, she continually utters this quick piping cry, and is immediately answered by the smothered cries of the imprisoned ones, who are just as anxious as she for the fray. If the swarming-fever is not yet allayed in the hive, this war-cry is bandied to and fro unceasingly; and the general ferment deepens, until, the condition of things having seemingly grown intolerable, the young queen rushes out, followed by the greater number of the bees. In the case of after-swarms, the concensus of evidence is in favour of the belief that the queen is really the leader of the party, although here again no positive rule is observed.

It may happen, however, that the stock is sick of all the turbulence and unrest that have so long beset it, and that the general desire is to restore the *status quo*. Under these conditions the sounds from the hive may have a very different quality and meaning. The queen still sends forth her shrill challenge, but now her cry is immediately followed by a curious hissing sound from the bees. It is exactly as if they were shouting her down, compelling her to silence by their own uproar; and when the war-cry of the first liberated queen is thus met by a chorus of disapprobation, it seldom happens that the stock swarms again. In a few days the queen goes forth alone on her honeymooning adventures; and on her return she is allowed to indulge her penchant for sororicide to her heart's content.

CHAPTER XII

THE COMB-BUILDERS

IN the foregoing chapters an attempt has been made to show that the honey-bee lives and moves and has her being in a world which must be actuated by something better than mere instinct, in the common usage of the term. To the modern biologist—the earnest out-of-door student of life under all its manifestations—this may appear as a rather obvious and unnecessary gilding of gold, and the only question yet undecided may seem to be where in the scale of reason the honey-bee is to find her equitable place.

All bee-lovers must plead guilty to an inveterate partizanship, the writer frankly among their number. There is no laodiceanism in bee-craft; and, all the world over, it may be said that, where a few bee-hives have been got together, there is always to be found a red-hot enthusiast not far off. The word "freemasonry," in the English tongue, has grown to be a synonym for the truest fraternity; but just as real, and almost as far-reaching, is the brotherhood among keepers of bees. No doubt, among themselves the tendency is rather to magnify the virtues and achievements of their charges: to be over-lavish of inference from too scanty or too isolated facts. And the proved impossibility of having anything to do with the honey-bee without being carried away sooner or later on a high wave of enthusiasm, makes any attempt at holding the balances truly between the zealous bee-lover and the interested but temperate-minded reader, a difficult and delicate task. Any writer on the honey-bee nowadays must be reckoned an ultra-specialist in an age of specialism; and here it is not easy to preserve the sense of proportion undimmed, especially for one admittedly speaking out of the ranks of beemanship, where all are aiders and abettors in ardour, impatient of any estimation falling short of high-water mark.

The story of the Comb-Builders, however, sets none of the usual pitfalls in the way of the over-enthusiastic penman. In its soberest incident and least important detail it is so wonderful, that exuberance of language is as powerless to exaggerate, as a niggardly tongue to minimise, its true and due effect. If the ordering of the bee-commonwealth—the intricate systems of sanitation, division of labour, treatment of the queen and worker-larvæ, and the like—is subject for marvel, and seems infallibly to denote the possession of high faculties, a much greater degree of acumen must be conceded to the worker-bee, when we come to consider her as the designer and builder of honeycomb.

It is here that she shines in her most significant light. The complicated structures with which she fills the bee-city do not call for unwearying toil alone: they could never have been fashioned unless the combined arts of engineer, architect, and mathematician had been brought to bear on them. Nor are they merely simple constructive and mathematical problems which the honey-bee is called upon to face; nor, though difficult, unvarying, and so amenable to instinctive solution. In almost every comb built we see special and necessarily unforeseen difficulties met and triumphantly overcome. In the construction of the six-sided cell, with its base composed of three rhombs or diamonds, the bee has adopted a form which our greatest arithmeticians admit to be the best possible for her requirements, and she endeavours to keep to this form wherever practicable. But it constantly happens, in her work of comb-building, that local conditions interfere with her plans; and then she will make five-sided cells, or square cells, or triangular, or any other form, just as the need impels her. It is a facile, comfortably finite thing to put all this down to a mysterious essence called instinct, with which the organism of the bee has been divinely dosed, as men serve electricity to a leyden-jar. But it was not instinct that made Wren put the steel cable round the dome of St. Paul's, nor instinct that lifted the crown-stones to the top of the Great Pyramids. These are works of a creature more highly equipped and instigated; yet their supremacy is all of a piece with the honey-comb, which is made of a material fragile, light as air, but which, by the art of the bee, becomes capable not only of supporting, but of *suspending* a weight thirty times as great as its own.

That the bee does not collect her building materials, but derives them from her own body, is a fact that has come to light only within the last hundred and fifty years or so, although several shrewd guesses at the truth are to be found in the works of the mediæval bee-masters. The wasp, who has much of the ingenuity of the honey-bee, but is doomed to exercise it in a far more humble direction, makes a six-sided cell; but her matter is collected from outside, and can only be put to comparatively simple uses, as it is incapable of bearing tensile strain. Beeswax alone, of all constructive materials in the world, seems to meet every requirement. It can be worked into plates as thin as the 1/180th part of an inch, which is the normal thickness of the cell-wall. It is indestructible to all the elements save heat. It can be rendered soft and easily workable, or allowed to harden, while still retaining its suppleness and life. It is a bad conductor of heat, and therefore conserves the heat of the hive. Vermin do not prey upon it: so far as is known there is only one creature that will eat it—a peculiar kind of moth-larva, against which, however, a strong stock can always hold its own. And then, as the raw materials for its production are secretions of the bee's own body, the work of preparing it can be carried on when darkness or stress of weather have put an end, for the time being, to work out of doors.

The first labour undertaken by a swarm, directly it has gained possession of its new quarters, is the building of combs. The apparent revulsion of feeling which succeeds the excitement of swarming soon passes off, and the energies of the whole party are at once concentrated on furnishing and victualling the new hive. The older bees commence foraging, each bee as she goes forth hovering a moment with her head towards the hive, to fix its location and appearance in her memory. By far the greater portion, however, remain at home and unite in a dense cluster for wax-making. Time is everything in these first operations of the new colony. The queen, with whom egg-laying has probably been suspended for a day past, or even longer, is overburdened with fecundity, and must be supplied with thousands of brood-cells without delay. The foragers will be coming home laden with nectar and pollen, and will need instant storage-room. Wax must be made with all possible expedition, and the young bees crowd together in the roof of the hive, with their queen snug and warm in their midst.

No doubt one of the chief reasons why swarming bees unite themselves in the solid pendant mass of the cluster so soon after leaving the parent-hive, is to hasten this process of wax-formation. It has been proved that wax is most easily generated under the influence of great heat, and this is well secured in the heart of the cluster. By the time the scouts have decided on the new home, and the swarm must rise again on the wing, a great number of the bees will have their wax-pockets filled, and will be ready for the work of comb-making. When a swarm is hived, even if it be only a short time after its issue, the little white wax-scales can be seen protruding from the armour-joints of many of the bees, and these are often dropped and lost in the general confusion.

One of the most difficult things to observe in bee-life is the actual process of comb-building. The crush is so great, and the movement of the bees so incessant, that at first the comb seems to grow of itself rather than be made by the busy multitude, for ever obscuring it from the watcher's eyes, or giving him but the rarest glimpse now and then of its white, delicate frailty of pattern. These early efforts of the comb-builders, produced as they are under forced circumstances, are occasionally faulty of design, as though hastily knocked together. Sometimes the first groups of cells made by a swarm will have a yellow, moist, spongy appearance, with thick, irregular walls, and are obviously little more than temporary vats to hold the incoming nectar until the proper honey-cells can be constructed. This emergency-comb is specially interesting, as affording one more instance of the worker-bee's ever-ready resource in the presence of difficulties. In the ordinary way the mason-bee hangs quietly in the cluster until her wax-secreting organs have done their work, and the six little oblong scales of brittle material are ready for manipulation. These protrude from under the hard plates of her abdomen, three on each side, looking much like half-posted letters. At one of the knee-joints of her hind-leg she has a peculiar implement, of which there is not the slightest trace in the queen-bee. This is like a pair of nippers, but instead of two converging points, it is furnished on one side with a row of sharp, stiff bristles, and on the other with a shallow spoon. With this special tool the worker-bee grips the wax-scale, and draws it out of its pocket. It is then transferred to her jaws, and she hurries off with it to the comb-building. Arrived at an unfinished cell, she sets to work to chew up the raw wax into a paste, incorporating it with her saliva, and materially increasing its bulk. The resulting soft, ductile matter is then applied to the work, and moulded into its needed shape. In this way, with hundreds of workers going and coming, the delicate white fabric of brood and honey-comb is built up with extraordinary rapidity.

How the coarse, spongy comb, which swarms will sometimes manufacture, is produced cannot be definitely stated. It has all the appearance of having been made from raw wax, hurriedly masticated and kneaded up with honey, and probably this is its actual composition. The secretion from the salivary gland, is necessarily slow, and with time pressing and a horde of impatient foragers dinning about her ears, eager to unload and be off again to the clover, the ingenious mason-bee appears to have hit on the idea of using the contents of her honey-sac as a substitute. Nothing, however, but a mechanical admixture can take place between honey and the raw wax. This dissolves only under the influence of the bee's saliva, which has intensely acid properties.

To understand all that the bees have accomplished when a new empty hive has been filled throughout with waxen comb, it is necessary to follow the operations of the swarm pretty closely during the first few weeks of its separate life. It is a big undertaking, the building of an entire, new bee-city, and the problems that confront the builders are many and complicated. In the first place, whether she ever attains it or not, the worker-bee will aim at nothing short of perfection. Hereditary experience tells her exactly what are the home-requirements of the colony, and she now sets to work to fulfil them in the best imaginable way.

A city is to be built which is to accommodate twenty or thirty thousand individuals. Vast nursery-quarters must be constructed, as there may be as many as ten or twelve thousand youngsters to cradle at one and the same time. For at least six months of the year no food will be obtainable from outside, so that the city must contain large storehouses capable of holding more than a six months' supply. As the temperature in winter can be kept up only by the bodily warmth of the inhabitants, life in the city must be concentrated into the smallest possible space; and the materials of which the city is built must be heat-conserving, while its construction must allow of perfect ventilation at all times, and in summer it must permit a free circulation of air, that the surplus heat can be readily carried off. The city must be a fortress as well as a home, and be closed in on every side as a protection against its many enemies, as well as the weather.

There is another, and just as vital a condition governing its construction—the necessity for strict economy in material. If there were any natural substance having the qualities of tenacity, lightness, ductility, and strength which the bees could obtain out of doors instead of wax, no doubt they would use it for comb-building, and they would not spend hours of precious time and consume large quantities of hard-won stores in the manufacture of their own material. But it seems there is nothing in nature possessing the needful properties. Bees collect a resinous substance, notably from the buds of the poplar, which they use for stopping up crevices. They dilute this also into a varnish, with which they paint the finished combs, and sometimes even combine it with wax to form a rough filling; but it appears to be useless in cell-construction. The whole city must needs be made of wax, and wax alone; and the bees are as careful of this precious substance as a miser of his gold.

Starting with these conditions—efficient house-accommodation for the colony secured at the least cost in time, labour, and material—the bee tackles the problem before her with an ingenuity that is little short of astounding. She appears to begin with the central dominant unit of the difficulty, and to work outward, vanquishing subsidiary problems as she goes. Her line of reasoning seems to run somewhat in this way. To raise the young, and store the honey, there is needed some kind of cell or receptacle. The young larvæ being cylindrical in form, a cylindrical cell is indicated; and this shape will serve also for the honey-barrels. Not a few, however, but many thousands of these vessels will be required: they must therefore be placed close together, as well for economy of space as for natural warmth. The cells could be grouped together mouth upwards in horizontal planes, storey above storey; but such a method of construction would be economically unsound. To prevent sagging in the heat of the hive, and under the weight they will be called to bear, the cell-bases would have to be thickened collectively into a substantial floor, which would need shoring-up at intervals—after the manner of the wasps. But in this, much valuable material would be diverted from its proper use. Obviously, a better plan would be to lay all the cells on their sides, and pile them up into a vertical wall. And, just as obviously, if two walls of these superimposed cells were placed back to back, so that one central vertical sheet of wax would serve to stop the ends of all the cells, right and left, a saving of half the material used for the cell-bottoms would at once be effected.

But, so far, the design is still only in its crude, initial stage. The upright comb, consisting of a double pile of round cells, back to back, with one flat base between, although a great advance on the single sheet of horizontal cells, is yet mechanically and economically deficient. The round cells leave useless interstices, which take much wax in the filling; while the flat bottoms do not coincide with the form of the larvæ, and thus still more space is wasted. Clearly, improvement can only come by altering the shape of the cell; and now the bee seems to have asked herself—and triumphantly answered—an extremely complex question.

She knew how much internal cell-space each larva required for growth. The problem, therefore, was this: of what shape, nearly approaching the cylindrical, ought such a cell to be made, which would ensure the right dimensions, but which would occupy the least possible room, have the greatest possible strength, consume the least possible material in its manufacture, and possess the property that a number of similar cells could be built up in a double vertical plane, leaving no interstices either between the cells or between the planes?

There is only one solution to this problem; and the honey-bee found it—who shall say how many ages ago?—in the hexagon cell, with its base composed of three rhombs.

The whole astounding ingenuity of the thing can only be realised when a piece of nearly perfect, new-made, virgin-comb has been closely examined. It will be at once seen that the hexagon cells combine together over the surface of the comb in absolute geometrical union, and that the six-sided form is round enough for all practical purposes. Looking into the cells on one side of the comb, it will be noted that their bases take the form of depressed pyramids, each made up of three diamond-shaped planes. Turning the comb over, we see that the cells on this side also have pyramidal bottoms. If the depth of a cell on one side of the comb be taken, and added to the depth of a cell on the other side, and then the width of the whole comb be measured, it will be found that the combined depth of the two cells perceptibly exceeds the width of the whole comb. At first glance this seems like a case of the less including the greater, which is a manifest impossibility. But, holding the comb up to the light, a further discovery is made, and the seeming paradox is eliminated. The bottoms of the cells are so thin as to be almost transparent, and it is at once seen that the cells are not built end to end, in line, but that each cell-base on one side of the comb covers part of three cell-bases on the other. If the three diamonds, composing between them the triangular base of a single cell, be perforated with a needle, and the comb turned over, it will be found that the three perforations come each in a separate cell. Thus the saving in the total width of the comb is effected by allowing the pyramidal bases on each side to engage alternately like the teeth of a trap; instead of meeting point-blank, they overlap each other, and the faces of the pyramids are so contrived that each of them helps to close two cells.

There is another advantage in this arrangement which will be immediately obvious. The apex and three ribs of each pyramidal cell-base form foundation-lines for the cell-walls on the other side of the comb. This means that not only do all cell-walls abut on an arch, but that every cell-base is strengthened throughout by a triple girdering. The result is that the amount of wax required in the construction of the comb can be everywhere reduced to an absolute minimum. It becomes merely a question of what thickness of wax will retain the honey; and this experience proves to be no more than 1/180 part of an inch. The whole thing, indeed, might very well be taken as an ideal exemplar of the triumph of mind over matter.

The geometric principles brought into play in the construction of honey-comb have been a favourite study with mathematicians of all ages, and especially this rhombiform method adopted by the bee in flooring her cells. The rhomb is best described as a plane-figure whose four sides are equal, like those of a square, but whose angles are not right angles. In such a figure there are necessarily two greater angles and two smaller, facing each other in pairs. The three rhombs composing the base of the honey-cell lean together, as has been seen, in the form of a blunt pyramid; and—treating all angles as negligible factors—the bluntness of this pyramid is found to coincide very aptly with the shape of the full-grown larvæ. But this is not the only reason for the particular inclination given by the bee to the rhombs forming the base of each cell. Economy rules here, as in everything else she undertakes; and the truth that she has chosen the one and only form of cell-base which takes the least possible material to construct has received very striking confirmation.

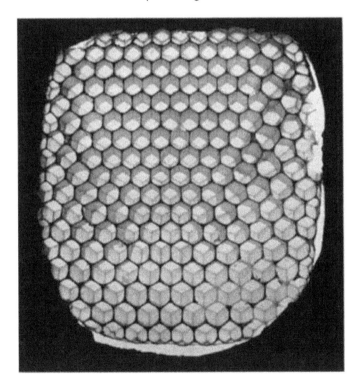

*Honey-Comb by Transmitted Light,
Showing Arrangement of Cells on Both Sides*

The story is an old and famous one, but it will bear repeating. A great naturalist once put himself to an infinity of trouble in measuring the angles formed by the rhombs in a vast number of comb-cell bases, and he found that these showed remarkable uniformity. It will be clear that the hollow pyramid of the cell-bottom will be either deep or shallow, according to the shape of the three rhombs composing it. The apex of the pyramid is formed by the meeting of three equal angles, one from each rhomb; and it is plain that this apex will be sharp or blunt, according to whether the meeting angles are wide or narrow. It was, of course, impossible to ascertain the dimensions of these angles with absolutely microscopical nicety; but, dealing only with the most perfect comb, the naturalist found that the two greater angles in the rhombs measured very nearly 110°, and the two lesser angles 70°. He also found that the angles formed by the conjunction of the cell-sides with the bases had the same dimensions as those of the rhombs. Assuming therefore that, mathematically, the angles of the rhombs and cell-sides should be equal, he was able to calculate exactly the angles for which the bees were evidently striving in the construction of the rhombs—109° 28" and 70° 32".

Another bee-lover scientist, ruminating over these figures, was much impressed by them, and determined to find out the reason why the bee made such constant choice of this particular shape of rhomb. He therefore conceived the idea of submitting the bee's judgment on this cell-base question to an independent authority. Without disclosing his object, he propounded the following problem to one of the greatest mathematicians of the day.

"Supposing," said he, in effect, "you were required to close the end of an hexagonal vessel by three rhombs or diamond-shaped plates, what angles must be given to these rhombs so that the greatest amount of space would be enclosed by the least amount of material?"

It was a difficult problem, but the mathematician worked it out at last, and his answer was "109° 26" and 70° 34"."

Now, the difference between the calculation of the man and the calculation of the bee was an exceedingly small one. No one thought of calling into question the work of the man, who was preeminent in his world of figures. It was therefore accepted as a fact that the bee had made a trifling mistake—so trifling, however, that, in the matter of comb-building, it was of no importance. Her reputation was unimpaired: to all intents and purposes the honey-cell was still a perfect example of utmost capacity secured by least material.

But another mathematician—a Scotsman this time—went over the whole business again, and he proved conclusively that the bee was right, while the first mathematician was wrong. He showed that the true answer to the problem of the angles was 109° 28" and 70° 32"—identically the figures obtained by estimation of the honey-comb.

In the foregoing pages the principles involved in the construction of honey-comb have been gone into rather minutely, because it is here that the lines of thought between the old and the new naturalists seem to make a typical divergence. Both schools are, in the main, agreed on the point that all forms of life emanate from the one omnipotent source; and it matters little whether we speak of the vast periods of time, during which the creation of all things was effected, as ages, or under the old Biblical metaphor of days. But whereas the old school appears to insist on different qualities of life—immortal soul in man, and a mystic, subconscious, perishable thing called instinct in the brute creation—the new school is unable to see any distinction between the intellectual equipment of man and brute, but that of degree. Between the honey-bee and her masters there is indeed a great gulf fixed, but it is conceivably not unbridgable. And unless we are determined at all cost of logical violence to force a favourite set of square opinions into the round holes of observed fact, it is difficult to see how the old position is long to remain tenable.

With regard to this particular question of comb-building,

an attempt is still being made to show that it is entirely due to the working of certain natural laws, and is independent of any intelligence or volition which the bees are supposed to exercise. We are told that the cells are always begun in a circular form, but that they afterwards assume the hexagon shape quite automatically, in obedience to the laws of mutual interference and pressure. As a proof of this, it is pointed out that the outside cells of the comb, not being subject to these laws, are usually more or less rounded.

The pressure-theory is hardly worth serious consideration, as it is obvious that the growth of a honey-comb is perfectly free and untrammelled in every way. If the bee makes her comb-cells with six sides and a pyramidal base unthinkingly, and under the yoke of imperious obligation, it is certainly not because the cells force this shape upon one another, like Buffon's peas in a bottle.

And if we believe that the bee works blindly under the law of mutual interference, any close examination of the results of her work must bring us to the conviction that we are only putting aside one marvel for something more wonderful still. For then we see a natural law taking on a very unnatural quality—that of intelligent adaptation to circumstances. The comb, intended for use in the hive-nursery, is made in two sizes. That used for cradling the worker-brood has cells measuring 1/2 inch across, and a fraction less than 1/2inch deep, while that designed for raising the drone-larvæ is built up of cells having a diameter of 1/4inch, and a depth of about 5/8inch. These different-sized cells are not mingled indiscriminately over the comb, but are grouped together in large blocks. Some of the combs will be entirely composed of worker-cells, which are always in the vast majority; other combs will be made up of both kinds.

The bees begin a comb by attaching a small block of wax to the roof of the hive. On either side of this they hollow out depressions, which become the bases of the first cells. The work is then extended downwards and sideways, the cell-bases being multiplied in all directions as fast as possible, so that there are a great number of unfinished cells in progress long before the walls of the first cells have been completed. There is a very reasonable motive for this procedure. When a house is being built, as much of the foundations as possible are laid in at the commencement, to allow a large body of bricklayers to get to work on the walls at the same time; and the bee extends her comb-foundations on the same principle.

When about half the comb has been finished for worker-brood, it may be decided to commence building drone-cells. As the bases of the drone-cells are larger than those of the worker-cells, it follows that a change must be effected in the ground-plan of the comb. The bees prepare for this transition very cleverly, evidently studying how the regularity of the comb may be least interrupted. Sometimes the change is contrived without any appreciable loss of space, but more often several misshapen cells have to be made before the symmetrical progress of the comb is resumed. This depends largely on the inherited skill of the bees, which varies according to their strain, as all ex perienced bee-keepers know.

Now, if the work of comb-building is carried through by the bees under blind compulsion of the natural laws of mutual interference and pressure, what other law, it may be asked, interferes with these in turn when the transition from one size of cell to another must be made? If it is all a sort of crystallisation going on independently of the bees' will or wish, it appears more than curious that the mill should grind large or small, just as the needs of the hive demand it.

But the whole position is really little else than a flagrant example of the evils of argument from a simile. Soaked peas in a bottle will swell to hexagons, or rather, dodecahedrons, by the law of mutual interference. Soap-bubbles will do the same with no more constriction than their own weight. But peas and bubbles are things self-contained and separately existing, before being brought together. If the bees made a vast number of separate, round cells, and then combined them simultaneously, no doubt all but the outside cells would assume the hexagon form. But the essence of the whole art and ingenuity of comb-building lies in the fact that there is no such thing as a separate cell. Each single compartment in the comb shares its parts with no less than nine other compartments. And to talk of mutual interference when there is no separate existence is ploughing the sands indeed.

There are other circumstances connected with the work of the comb-builders which go far to confirm the position that bees do exercise reason, and that of a high order. It has been said that the interior of a hive in day-time is not altogether deprived of light. Probably, during the hours of greatest activity, the bees have always enough light to see their way about by means of their wonderful indoor-eyes, which, under the microscope, have all the solemn wisdom of an owl's. It is a fact, however, that comb-building is usually carried on at night-time, when other employments are in temporary abeyance. Possibly the—to our eyes—profoundest darkness may be no darkness at all to the bees; but, to all appearances, as we can judge of them, honey-comb is virtually made in the dark.

But combs are built side by side, often simultaneously. They grow downwards together, yet always preserve their right distance apart; so that, when finished, there will be an intervening gangway between the sealed surfaces of about a quarter of an inch, which is just enough to allow the two streams of bees to pass each other, back to back. How are these distances preserved, seeing that the bees at work on the bottom edge of each comb are separated by a space of, perhaps, an inch and a half of empty darkness?

A simple experiment will at once give a clue to this. If a hive, in which a swarm has constructed about half its depth of comb, be canted a little sideways, so as to throw the combs out of the perpendicular, and the hive be then left for several days, it will be found on examination that all building, from the moment of disturbance, has followed on the new line of verticality. The combs will all be slightly bent to one side. This means either that the bees have a natural sense of the perpendicular, or that they work by the plumb-line, as humanity is constrained to do. The fact seems to be that the hanging cluster of wax-making bees performs the office of a living plummet, and really guides the comb in its downward progress.

Yet, do bees always suspend their combs? Do they never construct a waxen storehouse, raising it tier above tier from the floor of the hive, after the system of the more intelligent creature, Man?

The first commentary on this is, that such a departure from their common methods would be no improvement, but a retrograde step. These long comb-walls of the bees have a close analogy to the modern transatlantic sky-scraper building. The trouble with all such buildings is to provide them with sufficient base for their height. If American engineers had at their disposal a material of adequate tensile strength, and there were anything in nature to hang them from, it would be, scientifically, a better plan to suspend these buildings than to erect them, because the house would then naturally tend to keep its verticality, and the base-problem would cease to exist. On the same principle the bees, having at hand a material of almost ideal tensility, and a suitable hanging-beam, wisely suspend their heavily weighted combs from the roof, instead of erecting them, like certain kinds of ant-structures.

But it is undoubtedly long racial experience, and not inability to follow the humanly approved method, that guides them here. Rarely—so rarely that the writer, in the course of many years spent among bees, has seen only a single example of it—bees will build comb *upwards*, if circumstances will allow no other way. And this would seem not only to drive the last coffin-nail for the poor instinct-theory, but to carve its epitaph as well.

In the instance referred to, a glass-bottomed box had been inverted over the feed-hole of a common hive, and had there remained forgotten.

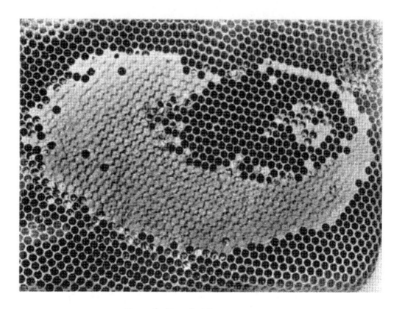

*Brood-Comb Showing Eggs,
Larvæ in Different Stages of Growth,
Sealed Cells, and Young Bees Cutting Their Way Out*

As the season progressed, the hive grew great with bees and honey, and it became imperative to build additional store-comb in the box overhead. But its slippery glass roof would give no foothold to the builders. Time and again they must have tried to get upon it, with their wax-hods filled and ready, and each time failed: the ordinary way of comb-building was clearly impossible. Then the engineers of the hive, inspired by the difficulty, got to work in another way. On the wooden surface below they laid out the plan of a garner-house, not after their usual method of parallel combs, but a regular, oblong house, with cellular storerooms, and communicating passages in between. Upon this they raised storey above storey of horizontal cells, until the glass roof was nearly reached. At this stage, apparently, the honey-flow came to an end in the fields, for the cells in the store-house were never sealed, though all were nearly full of honey; and later in the season it was found and carried away by the bee-master, who still preserves it as a curiosity. He bears a well-known name,[1] and his testimony as to the making of this unique little honey-house is beyond question; but, indeed, it carries in itself infallible evidence of its authenticity. All honey-cells made by bees have a slight upward inclination, which helps, as has been already explained, to retain their contents until they can be capped over. And every cell in the storehouse clearly showed this upward slant.

Footnote:

[1] Dr. Herbert MacDonald Phillpotts, of Kingswear, Devon.

CHAPTER XIII

WHERE THE BEE SUCKS

IT is characteristic of those unlettered in bee-craft that they are often afraid when there is no danger, and will venture with the intrepidity that is born of ignorance where old experienced bee-keepers fear to tread.

Temper in bees is one of the most variable qualities in a creature made up of variabilities. There are times, when a summer storm is threatening and the air is charged with electricity, when to go among the bees is to court certain disaster; and there are other times, such as the full height of the honey-flow, when almost any liberties can be taken with bees, without fear of reprisals. And yet this is not always the rule. Much depends on their lineage and the purity of the strain, and, again, on the systems of the bee-master. Bees respond as readily as any other form of domestic stock to wise and considerate treatment. Handled in a firm, quiet, deliberate way, the most vicious colony can often be dealt with in perfect safety; while the mildest-natured bees will commonly meet fumbling indexterity with a prompt challenge to war.

Since the Italian bee was brought to England, some half-century ago, there is no doubt that the original English strain has been greatly modified. Some authorities, indeed, question whether there are any absolutely pure British bees left at all. The golden girdles of the Italian crop up in the most unlikely places, and the foreign blood seems to have got into the race in all but the remotest parts of the country. One must regret, although it is a vain regret now, that these undesirable aliens were ever allowed to set foot on the soil. Whatever naturally survives and thrives in a particular country, must be the most suitable thing for that country; and these southern races of the honey-bee seem to have brought back, to the detriment of our own stock, idiosyncrasies long ago bred out of the native race. Much of the nervous irritability and proneness to disease visible in the honey-bee of to-day is more or less directly traceable to the introduction of foreign blood; and the grand special advantage of the Italian bee—its much vaunted and widely advertised possession of a long tongue—has proved an entire myth. Numberless measurements undertaken by our leading scientific apiarians have proved that the Italian bee has a tongue no longer than any other, although most are willing to concede her the possession of a very long and ready sting indeed. But here we do her an injustice: a pure-bred Italian worker-bee is as good or as bad tempered as any other of her species. It is the first crosses with the native bee which display so much vindictive aggressiveness, and have given to the whole race its general bad name.

In the time of the great honey-flow—which in southern England begins in May, early or late, according to the season, and may endure for six weeks—it is a common thing in the country to see people turn back from the footpaths, running through the white-clover or sainfoin fields, because of the huge and terrifying uproar made by the foraging bees. When there is a large acreage under these crops, and the day is a fair one, this note reaches a volume hardly to be credited as a sound of work and peace. It is much more like the din of a great bee-war, and it is small wonder that the stranger, unlearned in the ways of the hives, should fear to go through what is very like a scene of battle and carnage.

And yet there is no time of year when the honey-bee is so little inclined to molest her human fellow-creatures as this. So long as the honey-weather holds—the warm nights when the nectar is secreted, and the rainless days when it can be gathered—she can hardly be induced to attack, even if her home is being turned inside out, and the sudden sunlight riddling its darkness through and through.

Until within comparatively recent years it was universally believed that honey was a pure, untouched secretion from the flowers; and that beyond gathering and storing it the bee had no part in its production. This idea, however, is a wholly mistaken one. Honey is a manufactured article, and differs in almost every way from the raw juices obtained from the various flower-crops. The nectar of flowers, before collection by the bee, seems to have hardly any of the constituents of ripe honey. Three-quarters of its bulk consists of plain water, in which about 20 per cent, of cane-sugar is dissolved, the rest being made up of essential oils and gums, which give it its distinctive flavour. But mature honey contains very little water, certainly never more than a sixth part of its bulk. Its sugar is almost entirely grape-sugar. It is decidedly acid, while the nectar is always neutral. And the oils and aromatic principles of the flower-juices are matured and developed into the well-known honey flavour, which is like nothing else in the world.

It is certain that the process of manufacture begins directly the bee draws the nectar from the flower-cup. As the liquid passes into the honey-sac it is mingled with the acid secretion from the gland at the base of the tongue. When the bee reaches the hive she does not pour her burden direct into the cells, but passes it on to one of the house-bees, who conveys it to the honey-vats. It is even probable that the nectar is transferred a second time before it reaches the cell, although this point is still undecided. The effect of such transference is to add more acid properties to the original juice.

The honey seems to undergo a regular brewing process within the hive. It is kept at a temperature of about 80° or 85°, and it is then that the surplus water passes off into vapour. In this way the raw nectar loses at least two-thirds of its natural bulk before it is finally converted into honey. It is said that at the last moment, just before each cell is stopped with an impervious covering of wax, the bee turns herself about, and injects into the honey a drop of the poison from her sting; but there seems to be not the slightest evidence in support of this. The contents of the poison-sac are, it is true, mainly formic acid, which is a strong preservative; and undoubtedly traces of formic acid are to be found in all honeys. It has been, however, conclusively proved that this acid finds its way into the honey from the glandular system of the bee, and not through its sting.

The industry of the bee in nectar-gathering has always been a stock subject for wonder, and it is commonly supposed that she is born with full instinctive capabilities for her task. A little observation, however, soon tends to upset this theory. The work of foraging has to be learnt step by step, like every other species of skilled work in hive-life. The young bee, setting out on her first flight, has all the will to do well, and her imitative faculty is strongly developed; but she seems to have very little else. Her first experiences are a succession of blunders. She appears not to know for certain where to look for the coveted sweets, and can be seen industriously searching the most unlikely places—crevices in walls, tufts of grass, or the leaves of a plant instead of its flowers. The fact that the nectar is hidden deep down in the cup of the flower, beyond its pollen-bearing mechanism, seems to dawn upon her only after much thought and many fruitless essays.

It has been proved that bees will go as far as two or even three miles in their foraging journeys. The distance seems to vary according to the nature of the country. Bees in hilly districts appear to venture only short distances from home, while in flat country the foraging flights are more extended A bee-line has become proverbial for a straight course, but it is doubtful whether the bee ever makes a perfectly direct flight from point to point. The truth seems to be that there are well-defined air-paths out from and home to every bee-garden, and that these are continually thronged with bees going and returning throughout the working hours of the day. These aërial thoroughfares lie high above all but the tallest obstacles, so high indeed that the keenest sight will reveal nothing. Only the busy song of the travellers can be heard, like a river of music, far overhead.

In the South Downcountry, where the isolated farms are each surrounded with their compact acreage of blossoming sheep-feed, and there is nothing but empty miles of close-cropped turf between, these bee-roads in the air can be easily found and studied. Walking over the springy, undulating grass in the quiet of a summer's morning, a faint, far-off note breaks suddenly upon you like the twang of a harp-string high up in the blue. A step or two onward and you lose it; retracing your path, it peals out again. You can see nothing, strain your eyes as you will; but its cause is evident, and with a little trying you can presently make out the main direction of the flight, and see down in the hollow far below, the huddled roofs of a farmstead with a patchwork of fields about it, white with clover, or rose-red with sainfoin in fullest bloom.

Perhaps there is no honey in the world so fine as that to be obtained from these solitary Downland settlements. With the ordinary consumer honey is merely honey, and there is an end of the matter. But the beeman knows that the quality of honey varies as greatly as that of wine. He will tell you at first taste the crop from which it is gathered, whether it has one source or many, whether it is all flower-essence, or has been contaminated by the hateful honeydew, which is not honey at all. Down in the lowlands, except at certain rare seasons when only one crop is in flower, it is next to impossible to get honey absolutely from a single source. But here on the hills the bees are not tempted by glowing gardens with their feeble, washy sweets; nor are they led aside by the coarse-natured privet, or horse-chestnut, or sunflower. There is only one trencher to their banquet, but this is a vast, illimitable one. They have nothing to do but to wend out and home all day long between their hives and a single field.

It is difficult to guage with anything like approximate truth the amount of honey that one flowering crop will yield. But probably, when all conditions are most favourable, every acre of Dutch clover will produce about five pounds of pure honey for each day it is left standing in full bloom. The nectar is obviously secreted by the flower as an attraction to the bee, who, blundering into it with her pollen-smothered body, unconsciously effects its fertilisation. Directly this object is gained, the flow of nectar in each particular floret appears to cease, and the bee passes it by.

The student of old books on apiculture is often surprised to read so much in praise of honeydew, while in the modern bee-garden he hears of it nothing but hearty condemnation. He is told that directly the bees begin to gather honeydew the store-racks must be removed from the hives, or the good honey will be ruined both in colour and flavour. He is shown some dark, ill-looking, watery stuff carefully sealed up by the bees, and is informed that it is nearly all honeydew. But, he asks himself, can this be the same thing about which the old masters were led into such ardent eulogy? The truth is that when ancient and mediæval writers spoke of honeydew, they used the word as a general term for all that the bees gathered. Honey was all a dew, divinely rained down from the skies; and it is entirely of a piece with the all but universal lack of bee-knowledge down almost to the beginning of the nineteenth century, that so few should have guessed that the flowers themselves had anything to do with the matter. Virgil and the rest of the classics held absolute sway over all minds pretending to the least culture, and even the naturalists seem to have studied the wild life around them with no other object than to force facts into line with ancient poetic fantasies. The old writers explained the varying qualities of honey as being due to the influence of whatever stars happened to be in the ascendant at the time of its gathering, and the honey was good or bad according to whether this was favourable or unfavourable.

The quality and consistency of honey varies extraordinarily as between the different sources of true nectar; but there is no doubt that honeydew well merits the evil name it has gained with modern bee-keepers. There are, perhaps, three hundred distinct kinds of aphides known to English naturalists, and all these eject the sweet liquid which, under certain conditions, bees are tempted to gather. This honeydew varies in flavour according to the species of tree from whose sap it is derived. Probably much of it is only a sweet, slightly mawkish liquor, which, in its pure state, combines with the genuine honey without causing noticeable deterioration, at least to the unexpert taste and eye. But, unfortunately for bee-keepers, the oak is a great favourite with these parasites, no fewer than six varieties preying on this one tree alone. And oak-honeydew is a pestilent thing indeed.

It is commonly supposed that the first cold nights, that mark the beginning of the end of the honey season, stimulate the production of honeydew; for it is after a chilly night that bees are usually seen at work on the trees where the aphides abound. A much more likely theory, however, is that the cold does not accelerate the secretion of the honeydew, but cuts off the more legitimate resources of the hive just when they are in fullest activity; and so the huge armies of foragers are momentarily thrown out of work, and must seek new outlets for their energy. The secretion of true nectar takes place mainly at night, and requires a temperature of about 70°. Anything much lower than this means dearth on the morrow, no matter how fine and warm the weather may then prove,

The dark colour of aphis-syrup—a very little of which will ruin for market the finest honey—seems to be due as much to foreign matter as to its natural evil character. There is a peculiar growth on the bark of many trees where aphides congregate, which is known as soot-fungus. This and the honeydew get mingled together in a cimmerian slime, and, no doubt, the merest trace of it would serve to darken and spoil the purest honey. There seems to be no way for bee-keepers but to watch for the first chilly nights, as the honey-season draws towards its close; and then to be up early and get the surplus honey-chambers off the hives, before the bees have had a chance to spoil them. But the bee is no desperately early riser, for all her lofty place in the moral-maxim books. She generally waits until the morning sun has drunk up the night dews, and warmed the flower-calyces, before getting down to her work in earnest. The very early bees that may sometimes be seen winging out into the first light of a summer's morning, are probably only water-carriers. The water-supply is the day's first and last care with each hive in the breeding season. Every bee-garden seems to have its regular watering-place, generally on the oozy margin of some neighbouring pond; and here, in the early morning, and again towards late afternoon, the bees may be seen drinking in whole battalions, while the meridian hours of the day will find it all but deserted. Curiously, these water-fetching times coincide with the times when the nectar is least get-atable, or when the supply is exhausted for the day; which is another sidelight on honey-bee economics.

To follow the bees through their honey-harvesting season is to review nearly the whole year's natural growth and life. In southern England the earliest nectar is drawn from the willows, which come into flower with late March, but hold back their sweets until the first spate of fine hot weather comes flooding in the track of the chilly northern gales. Of willow-honey there may be much or little, according to the night-temperatures. Generally it goes by fits and starts. For a day or two here and there the trees may be crowded with bees, or they may be deserted for weeks together. Whenever the sun shines, indeed, the trees that stand up like torches of gold in the misty purple of budding woods, are always full of the singing multitude; but these are only the pollen-gatherers. The nectar-bearing willows are far less showy. Their catkins are small, tight-girt tassels of green, and when a warm night has brought them into profit, they attract all the noisy minstrels for miles round. Bee-keepers generally seem to leave the willows out of their calculations as a source of honey, but in riverside districts, and in favourable seasons, they are not to be overlooked. It sometimes happens that April comes in with a succession of mild sunny days and warm nights, and then the hives may suddenly overflow with willow-honey. When the yellow catkins fade out of sight, the willows are apt to fade out of memory; and it does not seem to be commonly known that the female catkins continue to secrete abundant nectar often up to the end of May.

Good honey-years are scarce under the changing English skies; yet Nature's design for the hive-people is obviously to give an unbroken succession of honey-yielding plants throughout the whole spring and summer, and pollen whenever a bright break of sunshine may lure them out of doors. The white-clover is seldom ready until the first week in June; but, from the earliest willows in March until the last of the flowering seed-crops is down in late July, there is abundance of provender, if only the fickle sun will do its part in the matter.

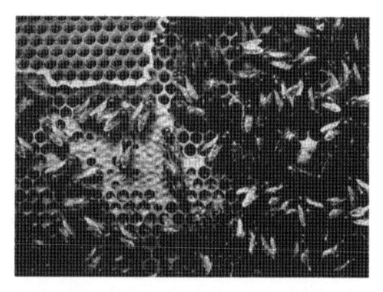

In the Store-House: Sealing up the New Honey

The clover, as farming goes nowadays, is the great main source of honey, in southern England at least; but the connoisseurs are at variance as to what yields the absolute perfection of honey.

Scotsmen are all of one mind, for a rare chance, in this; and will hear of nothing but the heather, carefully discriminating between the bell-heather, which is good, and the ling-heather, which is immeasurably better. Yet there is a honey, or rather a honey-blend, which far outstrips them all, though it is as rare and almost as priceless as the once famous Comet vintages. It is to be had only when the apple-blossom and the hawthorn come into full flower together, and this is only when a chill April has delayed the one and a summer-like May has forced on the other. Then, to the mellow refinement of the apple-nectar, is added the delicate almond flavour of the hawthorn, and the resulting honey is easily the finest sweetmeat in the world.

Wonder is often expressed that one of the most generally cultivated crops, the red-clover, is seldom visited by the honey-bee, although the bumble-bees fill it with their deep trombone-music at all times of the day. It is true that the tongue of the hive-bee cannot reach to the bottom of the long red-clover calyx, but this would not deter her it the nectar were worth the gathering. She would cut through the petal at its base, as she does with many other flowers, and so steal an effective march on her better caparisoned rival. But red-clover nectar is poor in consistency and coarse of flavour. When the main crop is in flower, it would yield a practically unlimited amount of honey, but this is just the time when the bee can employ herself more profitably elsewhere. After the red-clover has been cut, a second growth springs up, bearing flower-tubes less developed, and therefore shorter than those of the first crop. But now other and better sources of supply are rapidly failing. The bee—for whom, in prosperous times, nothing but the best is good enough—must revise her tastes to meet her necessities. At this time she is as busy as the rest in the red-clover fields. And when her clearer, sweeter note is heard there, mingling its contralto with the hoarser music of the bumblebee, it is a token that the heyday of the year is past: the honey-chambers must be taken off the hives without delay.

CHAPTER XIV

THE DRONE AND HIS STORY

IT is true that all bee-keepers are enthusiasts, and true that long years spent in the companionship of the hives invariably create a fearless fellowship, a prime understanding between the bee-master and his legions. But it is equally true that the longer you study the nature of the honey-bee, the less enamoured you become of certain of her ways.

In the minds of old bee-men there grows up, as the years glide, a sort of awe of her. She is so manifestly a power, supreme in her little world. She is so courageous, resourceful, brainy. All the weaknesses and compromises, and most of the pleasures, have long ago been driven out of her life, seemingly by her own act and will; yet, in doing this, she has but refined the science of citizenship to its pure elements. Her entire unselfishness, her readiness to sacrifice her individual good for the good of the State, are as unquestionable as they are changeless. The hive-polity, taken as a whole, is so admirable, and compares so advantageously with certain human efforts in the same way, that you are apt to exalt all her qualities into virtues; and to conclude that a far-seeing, wise benevolence must have gone to the making of the perfect Bee-State, instead of the cold, undeviating logic that alone has fashioned it.

This remorseless smelting-down of life into the set moulds of principle, without mercy and without reproach, has a cumulative effect on the mind of the observer; and sooner or later, though he will early lose his fear of her sting, he will develop a very real, but vague, awe of the honey-bee in another way.

Just as Moses Rusden, the King's bee-master, held up the life of the hive as Nature's evidence of the Divine will in earthly monarchy, so the latter-day student is often constrained to ask himself whether the bee-commonwealth does not point an authoritative moral in another way. Here is a State—only a mimic one, but still not a negligible example—where several of the most fiercely-debated questions of modern human life are seen in long adopted and perfected working-order, and seen in their fullness of result. Any attempt at a serious comparison between men and women, and the drone and worker-bee, would justly lay the writer open to the charge of grotesque trifling; but there is more than a fanciful analogy between the principles on which all civilisations must be based, whether they are insect or human. It cannot be denied that the communal life of the honey-bee is a high civilisation; that it has grown to be what it is to-day through ages of necessity; that the one sex has the other under a complete and terrible subjection, for which, and for the privilege of all power, the dominant sex has paid a terrible price.

The worker-bee to-day is an over-intellectual, neurotic, morbidly dutiful creature, while the drone is admittedly nothing but a stupid, happy, sensual lout. If the extreme difference between the sexes in bee-life had been aboriginal, the relations of drone and worker, as we see them in the hives to-day, would be meet and reasonable enough; but there seems to be clear evidence that, far back in the life of the race, the female bee was not so hopelessly superior to her mate. The queen-bee, in all likelihood, fairly represents the mother-bee as she was before the cooling crust of the earth made some sort of protected habitation necessary, which led first to close clustering for mutual warmth, and then gradually developed the complicated hive-life of to-day. But evolution will hardly account for all that we see: revolution must have had its part in the production of the modern self-unsexed worker. It has been seen that there is no physiological reason why each worker in the hive should not have grown into the fertile mother of thousands. The workers are not a stunted, specialised race, slowly evolved by time and necessity, and procreating their own stunted kind; but each worker is deliberately manufactured to a set pattern by the authorities in the hive, obedient to the call of the State. And when did the female bees begin this tampering with the springs of life, this improving upon Creation, which was the first vital step, failing which the present bee-commonwealth had been impossible? It looks very like a superb act of generalship in the great primæval war of sex—a brilliant piece of strategy that gave victory at a blow, and rendered the after-steps in the scheme of conquest a matter of logical sequence.

The whole question of the artificial production of the worker-bee is surrounded with difficulties; and it seems possible, on our present level of knowledge, to do little more than state the facts, and there leave them. The supremacy of the females in hive-life appears to have dated from the time that the vast majority deprived themselves, or were deprived by their immediate ancestors, of their share in procreation, and the ovipositor discovered itself as a weapon of offence and defence. Before the worker-bees existed as an armed force, there is no reason to suppose that the female bee had a great physical advantage over the drone. The queen-bee's propensity to thrust her ovipositor into the spiracles of her rival, and so effectually to despatch her, as well as her inveterate hatred of her kind, may both be late developments, due to the isolated, artificial life she now leads. While the worker is ever ready with her sting, the queen uses it so rarely that many old experienced beekeepers of the present time deny her altogether the power of stinging. A much more natural tendency with her is to bite; and when it comes to the use of the sharp, strong, sidelong jaws, the drone has a more redoubtable equipment than any, although he has apparently lost the will and sense to use it.

Whatever the drone may have been in far-off ages, the worker-bees have him now well under the iron heel of matriarchal expediency; and they see to it that he shall be fit only for the one indispensable office, although in that regard they exhaust every ingenuity to make him all that his kind should be. It is plain they would do without him altogether if that were possible. As it is, for nine months in the year there are no drones at all, and then only a few hundreds are raised in each hive—the bare minimum that will ensure the successful mating of the young queens when the summer sunshine calls them to their wooing. It might be supposed that where there are comparatively so few queens to be fertilised—only two or three at most from each hive, and these only once in a lifetime—that even those drones which are now tolerated are in excess of the number required. But a cardinal principle in bee-life is that the young queens shall choose their mates from another tribe, and so ensure a continual influx of new blood to the colony. This can only be effected out-of-doors, and as far as possible from the parent hive. The strongest impulse, therefore, of the virgin-queen, when she goes off on her mating-flight, is to get away quickly from her home surroundings. She flies straight off at tremendous speed, and thus has every chance of getting unperceived into new country, and so into the reconnoitring ground of strange drones.

Another reason for her extended flight and its remarkable pace is that only the strongest and swiftest drone of all the pursuing multitude is likely to overtake her, and this again makes for the betterment of the race. Perhaps there is no parallel instance in nature where the selection of the fittest individuals to continue a species is so carefully provided for, and no doubt this accounts for the high place of the honey-bee in the scale of created things. But this scheme involves enormous risk to the young queen. A hundred dangers lurk on her path. She is a tempting morsel for every bird that throngs the air of the June morning. Her untried wings may fail her. Even if she gets back safely to the bee-garden, she may enter the wrong hive, to her instant destruction. But she must take her chance of all risks; and the only thing to do is to render her absence from home as brief as may be, and her fertilisation as sure, by making the wandering-drone-population large enough to cover all probable ranges of flight.

From the very first the drone is nurtured in a different way from the worker-bee. The egg is laid in a wider and deeper cell; and during its first three days of life the drone-larva is fed with bee-milk, probably of a special kind and certainly of more generous quantity. After the third day this chyle-food is reduced, as is the case with the worker-grub; but while the worker is then given only honey, it is certain that the drone-larva receives both honey and pollen, and that for a full day longer. In all, it takes about twenty-four or twenty-five days to produce the perfect drone-bee, as against an average twenty-one days for the worker. The queen-bee, as has been already seen, is developed in much less time than either, little more than a fortnight elapsing between the time the egg is laid and the time she is ready to gnaw her way out of the cell.

After the drone is hatched, it will be another two weeks or so before he makes his first venture in the open air. All this time he has the free run of the larder, and steadily gorges himself on honey when he is not sleeping off the effects of his surfeit in some snug, out-of-the-way corner of the hive. But honey is not his only, or even his principal, food. Throughout his whole life he is constantly fed by the house-bees with the rich chyle-food given to him as a larva, and it has been proved that if this is withheld from him for the space of three days he will die of starvation, even in the midst of abundant honey. Thus the worker-bees have him completely in their power.

The first flight of the drones is a stirring event in the bee-garden. The common sound of the hives goes on practically the whole year through. Every sunny midday, when the temperature mounts to 45° or 50°, will see each hive the centre of a little galaxy of singers: it is only the volume of the music that varies with the waxing or waning days. But with the coming of the drones the whole symphony of the bee-garden abruptly changes. They never move from their snug indoor quarters until the day is wearing on towards noon, and then only in the brightest weather. Blundering aggressively through the crowd of busy foragers, they rise heavily on the wing, and soon the ordinary note of the garden is drowned in the new uproar. They seem to come almost simultaneously from all hives at once. For a minute or two the rich, hoarse melody holds the air; and then, almost as suddenly, it dies away, as these roystering ne'er-do-wells troop off over hill and dale, each to his favourite hunting-ground.

There is great divergence of opinion as to the limits of flight of the drone, but probably he goes farther and faster than any have yet credited. His magnificent stretch and strength of wing mark him for a flier. He is all brute force and lusty energy; and it would be strange if, with but one thing to do in life—to gad about in search of amorous adventure—he could not do it to a purpose. If a hive of bees be removed to a distance in the height of the season, some of both workers and drones are sure to find their way back to the old spot. This has constantly taken place when hives have been carried no farther than two miles. But in one case, when the distance was more than twice as much, no workers were seen round the old hive-station, yet a little company of drones was winging aimlessly about the tenantless stool, and there can be little doubt that these belonged to the removed colony. It is not suggested that they deliberately travelled all these miles. The chances are that, in their daily flight, they got so far away from the new station that they came within the zone of old landmarks, and thus naturally went on by the long-accustomed ways.

As a typical instance of a sluggard and idler, the drone-bee has enjoyed a vogue in the preparatory-school books for ages past. But, whatever his primæval equipment for usefulness may have been, it is evident now that he could not labour if he would. Physically, in all points but that of muscle, as well as mentally, he has become degraded to the inferior of the worker-bee in every way. He is destitute of all those special contrivances with which she is so amply furnished. He has no baskets for pollen-carrying, nor any of the ingenious brushes and combs which she uses to scrape the pollen from herself and others. He has neither wax-generating organs, nor leg-pincers to deal with wax. His tongue is too short for honey-getting. His brain is much smaller than even that of the feeble-minded queen. The intricate gland-systems, which play so important a rôle in the daily life of the worker, are either completely atrophied in the drone or exist only in an elementary state. While it has been the communal will of the hive that the worker-bee should develop an amazing proficiency of mind and body, the same forces have been steadily at work to degrade the male-bee into a creature of dependence, gradually training out of him all initiative and idea, except in the one direction. Just as in the case of the queen and the worker, drone and worker-bee seem hardly to belong to the same race.

And yet, for all his frank incapabilities and lack of ideals, the drone offers, in one respect, a refreshing contrast to his sour, stern, duty-worshipping sister. He is a life-long, incorrigible optimist. He fiddles gaily while the city burns. All his misery and mourning would not serve to quench a single spark of it; so he eats, drinks, and is merry, with the intuition of all drones that Nemesis waits on the morrow with something disagreeable. It is impossible to study his ways for long without recognising the spirit of rude jollity and horse-play that thoroughly pervades all he does. In and out of the hive he blusters, cannoning roughly against all he meets, and raising his burly, bullying song in the air as a sort of protest against all this anxious industry going on about him. Once gone from the neighbourhood of the hive, he seems to keep incessantly on the wing until hunger prompts him home again. For no one has ever seen a drone-bee among the insects that haunt the flowers, nor ever seen him basking on a sunlit wall or tree-trunk, after the kind of almost every other winged atom in the universe.

He comes back to the hive with the same noisy, careless fanfaronade, and is received by the workers with the same sullen indifference. They give him his fill of bee-milk, linking tongues with him as he sits up like an overgrown baby, voracious, clamouring to be fed. They suffer him to swill at the honey-stores unchecked, but plainly regard him with contumely. He is a terrible expense to the State, yet a necessary one. Silently they go about their uncongenial business of nourishing him—silently, and with an ominous patience. They grudge him every drop, and, all the more, urge him to his excesses. It is not for long. The day of reckoning is near at hand. Already the poppies glow scarlet on the hill—the poppies that mark the turning-point of the summer; and after them the long decline, with its ever-diminishing sun-glow; each day with a scantier meed of blossom, until the path runs again into the dreary levels, the sober greys and russets, of winter death.

Now the worker-bee is to show a grizzly seam in her nature, matching ill with the fine hues and qualities of mind for which she is so justly famed. And that she is not all lovable, all admirable, accounts for the exceeding love of her that moves the hearts of men who know her through and through. The story of the massacre of the drones has hardly a parallel for sheer relentless ferocity—unrecking abandonment to a vengeance long withheld for expediency's sake. There come the first chill nights of mid-July, and the honey-flow is suddenly at an end. The clover and sainfoin have already fallen to the sickle. Nothing but the bravest warmth and exuberance of the summer could now withstand the drain of the myriad honey-makers, and a few hours' cold dams up at once the attenuated stream. The time of prosperity is over There will be no more abundance of honey. It remains for the genius of hive-economy to prove how much of what has been gathered can be preserved for future needs.

The first sign of the *débâcle* is the throwing out at the hive entrance of certain pale, gruesome objects—the corpses of immature drones, not dead from mischance, but ruthlessly torn from their cells. This may go on intermittently for many days, and while the fell work is proceeding the living drones seem to take no warning. They keep up their merry round; the unending feast riots forward; daily the bee-garden is filled with their careless, overweening song. And then at last the signal for the slaughter is given. Within each hive a curious sobbing outcry begins—a cry that is nothing but sheer terror put into sound. The drones no longer lie in easy ranks between the combs, placidly sleeping off one debauch and dreaming of another. They are all awake now, and fleeing abjectly for their lives through the narrow ways of the bee-city, the workers in hot pursuit.

The deep, vibrant, horror-laden note increases hour by hour. As each executioner overtakes her victim, she grips him by the base of the wing; and, helped by others all alike infuriate at the work, she half drags, half pushes him through the throng, until she has him in the light of day, and tumbles with him to the ground; he for ever fighting and struggling, and uttering that frenzied note of fear; she savagely gnawing at the wing until it is disabled, and he can never more return to the hive. Many of the strongest drones escape from their persecutors for the time being, and fly away unhurt. But it is only for a few hours. Hunger is sure to bring them back to the hive, when the waiting guards fall upon them, and maim or drive them off once more. It is specially to be marked that the bees never sting the drones at this great annual feast of carnage. There is that much method in the madness which has seized upon them; for, in the rough-and-tumble of such a conflict, stings would be plucked out by the roots, and thus valuable lives would go down with the worthless. The sole object seems to be to rid the hives as effectively as possible of the presence of the drones; and the disablement of one wing appears to be all that is necessary, and therefore all for which the deft assassin strives.

With some bee-races the massacre of the drones is carried through in an incredibly short space of time; with others the agony of the thing is drawn out for days together. The wretched sires of the hive are caught between two evils, each as fatal as the other. If they fly off to the fields, starvation and the night-chills will swiftly bring about their end. If they return to the hive, a still speedier death awaits them. Night and day, at this time, the guard-bees are doubled and re-doubled at the city-gates; and there is little chance of the wiliest drone outwitting them. But he usually takes the home-hazard; and sooner or later comes blundering in, receiving with open arms, as it were, his share of the knife, as Huddlestone faced the Carbonari.

All this is the common way with the bee-republic, when the season goes as it should; and the hive is in possession of a mother-bee—young, strong, and of proved fecundity. But there are times when the drones—for all their great expense and drain on the wealth of the colony—are suffered to live on until the late autumn, or even to remain unmolested throughout the winter and following spring. If the bee-master sees drones about a hive, when other colonies have long ago made a good riddance of them, he well knows what ails the stock. Its queen is old and failing; and these astute amazons have given reprieve to their male-kind until a new mother-bee can be raised and properly mated. It is a case of mercy to the drones tempered with so much justice to themselves that the original virtue is largely discounted.

And where the drones are carried through the winter, it is ever a sign that the hive is not only without a queen, but never will contrive one, of their own race. Yet they know that, in the preservation of the drones, they have at least one indispensable element for their salvation, and—who shall gainsay it of the sovereign honey-bee?—perhaps they rely on the bee-master to guess their plight, and furnish them with another queen, in time to save his property from extinction.

CHAPTER XV

AFTER THE FEAST

AS the year grows in the bee-garden, so it goes, with all but imperceptible tread and tread. In southern England, after the seed-hay is down, there is little more for the bees to do but prepare their hives for the coming winter. The queen is slowly weaned from her absorption in egg-laying by a gradual change in food. Day by day she receives less of the mysterious bee-milk which was her urging and inspiration; day after day she finds herself the more constrained to slake her hunger at the open honey-cells with the common crowd. Every day sees fewer bee-children born to the hive, and every day sees more and more of the old workers—worn out with a short six weeks or so of summer toil—pass away in that inexplicable fashion, using, perchance, their last strength of wing to hie them to the traditional graveyard of their kind. What becomes of them all, not the wisest among bee-men knows; but it is certain that, as they lived by communal principle, in the same faith they die; and their last act may be the truly collective one—of removing their own bodies out of the way of harm to the cherished State.

With the waning months, the population of the hive decreases visibly, and, as their numbers fail, the temper of the bees suffers just as evident a change. Old bee-keepers know by sharp experience that early autumn is a time when vigilance well repays itself. For all life the season of autumn has its peculiar tests and trials of character; and this is especially true with regard to the honey-bee. Each strain of bees has its proclivities, good or bad, which are sure to come to the front at this season. And, more than any, bad qualities will show themselves, now that the rush of the year's work is over, and the common energy must take its course through an ever shallowing and straitening way.

To find rank dishonesty in a creature of so small account in creation as an insect, is rather startling to old-fashioned ideas; but it is nevertheless beyond dispute that some stocks of bees are prone to develop a tendency to house-breaking and robbery of their neighbour's goods during early autumn, and, in a lesser degree, when the first scanty supply of nectar begins in early spring.

Virgil, and almost all the classic writers, give stirring accounts of the frequent battles among bees in their day. We are told of vast conflicts taking place in mid-air, of the kings leading forth their hosts of warriors—the din of carnage—the wounded and dying falling like rain out of the blue of the summer sky. These descriptions have always been a great puzzle to modern students of bee-life, because nothing of the kind seems to take place at the present day. Each hive goes about its business, apparently in complete disregard of the existence of other hives. Neither at home, nor abroad in the fields, are reprisals ever witnessed among bees, whether singly or collectively. The most peaceable creature in the world is the honey-bee, except in the single case when her home is being wantonly assailed.

But in autumn frequent encounters take place between robber-bees and the hive they are attacking, and one is constrained to believe that it is of this Virgil writes.

Perhaps when once a stock has discovered that stealing honey is a much quicker and easier method of obtaining it than by the laborious process of gathering, these particular bees will never again be won back to honest courses. Not only will the parent hive continue to break out in this way at the close of every season, but all swarms from the same hive are certain to develop the like tendencies. The strain will be a continual source of annoyance and loss to the bee-master, and, if he be wise, he will take the shortest and surest way of putting an end to the trouble, by promptly changing the queen, and thus in the end exterminating the original stock. Where this is in his own garden, there will be no difficulty in the matter; but often the robbers are wild bees, brigands inhabiting a hollow tree in some neighbouring wood, and making sudden raids upon their law-abiding neighbours in adjacent villages, after the manner of brigands all the world over. The strangers have often a peculiar appearance, which singles them out immediately from the legitimate members of the gardens. They are darker in colour and shinier; and they have a bold, yet furtive, way of getting about, which suggests at once the prowling marauder.

Wandering among the hives on a fine September morning, several of these light-fingered, sinister folk may be seen hovering about the entrance to a hive, or trying to creep in unobserved. Their presence is promptly detected, and a sudden hubbub arises as the guard-bees set upon the intruders and drive them off. There is no doubt of their intention. They are spies from the robber camp, and their object is to discover those hives which are weak in population, and so will fall the easier prey to the depredators when in force. Strong stocks have little to fear from robbers; they can always hold their own against attack, and therefore are seldom molested.

These scouts disappear for a time, and the hive settles down to its wonted, busy tranquillity. But soon a little blur of bees may be seen coming over the hedge-top, and making straight for the selected hive. There is no more crafty reconnoitring. It is to be battle undisguised. The robbers descend upon their prey, and at once a terrific uproar begins, a desperate hand-to-hand fight between besiegers and besieged. Left to themselves, the weak stock will have little chance from the outset. It is quickly overcome. And then a curious thing often happens. The bees of the home-colony which have survived the fight, join forces with the victors, and themselves help to rifle and carry away to the robbers' lair the treasure which is their own by right. Luckily, the bee-master has an all but unfailing preventive of this vexatious trouble ready to his hand. He can safely leave all those hives which are numerically strong of citizens to take care of themselves, and those which are weak of population he can join together in twos or threes, converting them also into strong, self-protective colonies. The modern movable-comb hive is a power in the hands of the capable beeman, for the comb-frames from several hives can be placed together in one, and the bees will unite quite peaceably at this season, if all are well dusted with a flour-dredger, or treated with a scent-spray, so that in odour and appearance they may be alike. Probably every hive has its own distinct odour, which is shared by all its denizens, and this is no doubt the means by which the sentinel bees at the entrance recognise their own comrades, while they promptly fall upon all interloping strangers.

The preparation of the hive for the winter is of a piece with all else that the bee undertakes. As the area of the brood-nest shrinks, the empty cells are filled with honey, this being brought down from the store-cells farthest away. The foragers keep steadily at work whenever the weather holds, gathering up the remnants of the feast and bringing them home to swell the winter-larder. Where there is much ivy, a fine October will often see the hives as busy again as ever they were in the bravest days of June; but the throng of bees is manifestly smaller. The rich song of life begins later in the day, and lasts only during the brightest hours; and that wonderful night-sound, the deep underground thunder of the fanning bees, is gone from the bee-garden, just as the scent of the clover-nectar, brewing and steaming in the hives, no longer drifts across in the darkness, filling the bee-master's house with the fragrance he loves more than all else in the world.

The old ragged-winged bees, that have stood the brunt of the season, are now, too, nearly all gone. The hives are filled with bees of the same race, inspired by the same traditions; but they are at the beginning of life, the raw recruits of destiny, a mere stop-gap crew. They have no memories of the time when work was a fever, a tumultuous race with the sun, in which the swiftest must lag behind. They have never known the over-weighty cargoes, the bursting honey - sacs, and pollen-panniers so laden that they could be scarce dragged into the hive, and they will never know them. These bees, born late in the season, have their lot cast in the torpid backwaters of their little world. Theirs is to be but a dreary eking out of days, so that they may have strength enough to warm the first spring broods into life. The few hot days that burn in the midst of the snows of each English March—immeasurably far off now, and unattainable, seemingly—will be all they will ever see of the power of sunshine. Winter bees are born to the prison-house; and in it, and for it, live and die.

At the most, a worker-bee sees but six months of life: at the least—and this is the lot of many—she withstands the incessant wear and tear of her hard calling for six, or possibly eight, weeks. Thus, though the hive may be always packed with citizens, the population is for ever changing. Half a dozen times in the year, perhaps, and for a score of years, you may go to your bee-garden, and each time move among tens of thousands to whom you are an utter stranger, and whom you have never seen before. And yet, in all its customs, its propensities, its traditions, the life of the bees is Continuity impersonified. You may go round the world, and spend ten years on the journey; and, coming back to the old leafy nook of the country, find the old green hive still in its corner under the lilac, still the centre of what seems the same crowd of winged merchant-women sailing home under the same gay colours, singing the old glad songs, building the old wondrous fabrics in the darkness, transmuting the same fragrant essences into the same elixir of gold. And what is this mysterious thing called the Bee-Commonwealth, which is alone immortal, while all that composes it, and pertains to it, and upholds it, passes and dies?

You must not forget the queen-bee here. She alone, it must be remembered, persists year in and year out, while generation after generation of her children grow up and die about her—a hundred thousand of them, may-be, in each twelvemonth, thousands even between one single summer dawn and the dusk of the western sky. Methu-salah of old, on the more moderate human scale, must have had some such experience—must have divined the broader plan of life from the incessant repetitions of chance and change that passed before him. The power to generalise into symbols comes only to the ancient of days; and he of all men had learnt to fathom, to estimate, to winnow out the sober drab grain from the glittering, rainbow chaff of life. Over and over again he must have kept the true true to itself with one wise word, and turned back the false, dazzled and discomfited, with one flash from his mirror of the ages. He was a living history-book, where all men might read the common drift and outcome of life; and as a record of the hive's story, a living archive for its plans, its systems, its ideals, the mother-bee may exist to day—she who, in comparison with its ever coming and going thousands, is an age-old, imperishable thing.

And so you may think of her, in the short days of December twilight, or in the interminable night-darkness full of the raging of the winter wind, gathering her children about her, and telling them tales of their forbears' prowess; teaching them old bee-songs which have but the one refrain of work and winning; and never forgetting her own little story—of the one brief hour of her love-flight and marriage, bought and paid for by widowhood lasting her whole life.

CHAPTER XVI

THE MODERN BEE-FARM

IT is well enough to consider the scientific side of hive-life for its intrinsic interest, to treat it for what it really is—one of the most absorbing studies available for leisure hours. But the honey-bee is something more than a wonder-maker, or a peg on which to hang dilettante moralisms. Rightly treated and exactly understood, she can be made of great use in the world.

There are two things in this England of ours which profoundly astonish all who love bees, and have a true conception of their possibilities. Travel where you may in the land, the last thing you are likely to meet with is a bee-farm, or even a few hives in a cottage-garden; while every yard of your way has its nook of blossom, and every mile its stretch of flowery pasture, where, in sober truth, tons of honey are annually running to waste. All this could be garnered and sold to the people at little trouble and great profit, if only enterprise would wake up from its island-lethargy and stretch forth the hand. But the years dribble uselessly by, and nothing is done. Here and there a wide-awake husbandman gets a little township of hives together, sells in the neighbourhood all the honey his bees make, and puts to his pocket a gold and silver lining. But this is only a drop in the ocean, and the British people must send abroad for their honey, which they do to the pretty tune of more than £30,000 a year.

Hitherto, reasoning backward from effect to cause, it would seem that farming has been remunerative only when undertaken on a large scale; but those who can read the signs of the times tell us that the age, just dawning to the country-side, will be the age of the small man. And this must mean that the hereditary aristocracy among crops—wheat, oats, barley—will slowly give place to little-culture: in a word, that the land will be made to produce, not the things that tradition and our yeoman family pride have ordained as the be-all and end-all of farming, but the minor, humble necessities for which each town and village should look to the good brown earth immediately about it, but at present looks in vain. Farmers' ladies may then no longer sit in their drawing-rooms and ride in their carriages, but that will be a change for the simpler, more proportionate. Those who live in towns have little conception of it; but the country-dweller knows well what complexity and luxury have got into the old English farmhouses, for all the outcry about hard times; how the farmer's wife no longer goes to her dairy, nor makes any of the good old farmhouse things that served to uphold country England in days gone by; and how the master-agriculturists now are the sinews of the great London Stores, while the little local shopkeepers are left to the field-labourer with his twelve or fifteen shillings a week.

Chapman Honey-Plant in Village Garden

For the class of small-holders that must now multiply throughout the length and breadth of the land, there is awaiting an enterprise—a source of livelihood—as yet hardly tapped. A stock subject of envy with most artisans is the capitalist who leads an easy life while his factory hands toil for him. But if the small-holder will take up beekeeping, he too can look on, to a large extent, while his thousands of winged labourers are filling his storehouse with some of the most useful and saleable merchandise in the world. It is a truism in commerce that a good supply creates a demand just as certainly as that the universal want of a thing stimulates its production. One of the needs in England to-day is a full, good, and cheap supply of honey; and when this is forthcoming there will be little fear but that the present demand will increase hand over hand.

There are many reasons why the people should choose honey for their principal food rather than the beet sugar which is now so largely consumed. In the first place, honey is a pure, natural, undoctored sweet, while in the manufacture of ordinary sugar the use of more or less noxious chemicals seems to be indispensable. When a stock of bees must be artificially fed, and common grocers' sugar is used for the purpose, the result is generally that half the stock is poisoned by the chemicals with which the sugar has been treated at the mill. And if this is its effect on bees, the inference must be that it cannot prove altogether wholesome for men. But its purity is not the chief reason why honey should be the universal sweet-food of the people. Honey is the ordinary sugar of nectar concentrated and converted into what is chemically known as grape-sugar; and thus, in ripe honey, the first and most important part of digestion is already effected before it leaves the comb. This explains why so many delicate people, and particularly children, can assimilate food sweetened with honey, when they can take no other form of sweet.

Doctors are continually finding some new virtue n honey. Its gently regulating action has been long known, and there is good authority for stating that there is not an organ in the human body which does not benefit from its habitual use. In all wasting diseases, and triumphantly in consumption, it will prevail as an up-builder when everything else fails. There is no doubt at all that cases of consumption have been entirely cured by a liberal diet of honey; and, notoriously, honey is the main ingredient in nearly all patent medicines for diseases of the chest and throat. Therapeutic hints from laymen are generally looked upon askance by medical men—at least, by those of the old-fashioned type; yet, on the chance that this page may come under the eye of some of the more elastic-minded, the thing may be hazarded. There are many who believe in it, and with good reason, as a sovereign specific where the disease is a wasting one. It is nothing else than the once famous Athole Brose, which, as all Scottish bee-keepers know, consist of equal parts of good thick honey, preferably from ling-heather, and of cream, and of mature Scotch whisky from the pot-still. Little and often is the rule for its administration, but, unlike most old wife's remedies, faith has nothing to do with its wonder-working. Scepticism is a soil in which it seems to flourish as well as any.

The man of business, resolved to take up bee-keeping as a livelihood, must, at the outset, decide on what scale he will carry the matter through. There are two aspects of the thing, each more alluring than the other, according to the temperament and point of view. There is the Simple Life and the bee-garden — a life spent in the green quiet of an English village, within reach of a market town, where the produce of the hives may be disposed of. And there is the greater enterprise, the foundation of a bee-farm on an extensive scale, and on the most approved scientific principles, where the object is to supply the great central markets at a distance rather than the immediate local needs.

In the establishment of a bee-farm the first care must be the choice of a suitable district. The nature of the surrounding country must largely govern the systems on which the farm can be most profitably worked. The first maxim in successful beemanship is to get all hives filled to the brim with worker-bees by the time the great honey-flow sets in. This time, however, varies according to the district. In the orchard-country we need bees early; in heather-districts we want them late. In south-west England, where the country is half fruit-ground and half moorland, the hives must be huge in population both late and early. But where the bee-keeper follows the sheep-farmer—and there is no better guide to honey than the sheep—his true policy is to work his colonies slowly and steadily up to their greatest strength by the time the main feed-crops come into blossom, which is seldom before the middle of May. And all these considerations land us on the brink of a very vexed question in modern bee-craft—whether bees should be artificially fed, and if so, how and when?

If only the purest cane-sugar is used, and the syrup well boiled and never burnt, there is nothing to say against the practice on the score of harm to the stocks. Where early bees are wanted, it is absolutely necessary to give them a continuous supply of sugar-syrup from the first moment that breeding commences in the hives. Chemically, the sweet constituent in nectar is almost identical with that from the sugar-cane; and sugar-syrup has this advantage over honey given—that it more nearly simulates the natural flow. The bees responsible for the nursery-work in the hive and the regulation of the queen's fecundity, are young bees that have never yet flown. They can, therefore, only judge of the progress of the season by the amount of nectar and pollen coming into the hive. Where this is steadily increasing day by day—and it is this regular natural progress in prosperity which the bee-keeper must strive to imitate in artificial feeding — the nurse-bees gain confidence, and brood-raising forges rapidly ahead.

But sugar-syrup and pea-flour are not natural foods for bees, and there is little doubt that a prolonged course of such diet tends to lower the tone and stamina of the race, and thus may prepare the way for disease. The golden rule in the matter seems to be that artificial feeding should be resorted to only where strength of stocks is necessary to secure the harvest, or where actual starvation threatens. In purely heather-districts, when the big population is quite early enough if it is to hand in late June, nothing short of imminent starvation should induce the bee-master to give artificial, and therefore unavoidably inferior, food. In sheep-country the same rule holds. Except in the most unfavourable years, a hive, headed by a young and vigorous queen, can be relied upon to get itself into the finest fettle by the time the main crops are ready for exploitation. In this case the beeman has only to make certain from time to time that no stock is in absolute want of the ordinary means of subsistence.

But in those warm, favoured regions of the south-west, the lands of the apple-blossom and the heather, where there is a very early and a very late harvest to be gathered, a different system must be pursued. Here we touch on the second grand principle of successful bee-keeping—the necessity for having in all hives only the most prolific mother-bees. For profitable honey-getting a queen should seldom be kept beyond her second year. After that she is usually of little account, and should be superseded, either by the bee-master or the bees. But where a queen has been over-stimulated by feeding to raise an immense population in the spring of the year, she is rarely capable of another supreme effort in the autumn. The best policy, therefore, if the heather-harvest is an important one, is to remove the old queens as soon as the spring work is over, and to substitute for them queens that are in their best season, but at the beginning of their resources instead of at the end. In this way another huge army of workers is soon born to the hive, and the double harvest is secured.

On the question of the best hive to use in commercial bee-keeping, on either a large or small scale, it is hard to particularise. Generalisation, however, is not difficult here. Every bee-master has his own ideas as to details, but all are happily agreed on the main constructive principles. Experience has fairly well decided that a good queen, under the modern system of intensive culture, will require for her brood a comb-surface of about 1,800 square inches. A brood-nest of smaller capacity than this is liable to cramp her operations at their highest, and anything in excess of it will simply mean so much new honey lost to the super-chambers, where alone the bee-master requires it.

*Bad Beemanship:
Bees Lying-Out from a Too Crowded Hive*

Honey stored in the brood-nest, except during the off-season, is loss instead of gain. The best hive, therefore, will contain just as many brood-combs in movable frames as will ensure the right capacity; and all comb-frames throughout the bee-farm must be of the same size, so that they will be strictly interchangeable among the various hives. This is a vital point in successful bee-culture, because it enables the master not only to equalise the strength of his stocks by transferring combs of hatching brood from one to the other; but he can also give to penurious stocks frames of sealed honey from the abundance of their neighbours, and he can unite the weak colonies, thus rendering all strong.

For the rest, the hives must be so made that heat will be perfectly retained in the cold season, and as perfectly excluded during the sultriest time of year. Double walls round the brood-chamber are a necessity in the changeable British climate, where chilly days are always probable during ten months out of the twelve.

As well as honey-production, the bee-farmer will find an equal source of profit in the production of wax. Just as there is nothing like leather, beeswax holds its own as a marketable commodity in spite of paraffin substitutes. But if it is almost universally degraded by adulteration, the fault lies with the bee-men, who have never seriously attempted to meet the demand for it. Wax-production on a large scale is perfectly feasible, and there is little doubt that it could be developed into an important British industry, as it used to be in mediæval days. Yet these are times of revolution: the honey-bee may yet find herself entirely restored to her old national avocation—of bringing light to our darkness, and to our bodies one of the best and purest of foods.

CHAPTER XVII

BEE-KEEPING AND THE SIMPLE LIFE

IT is a quality of English sunshine that it comes and goes capriciously, so that no man may be sure of the comradeship of his shadow from day to day. But when there is sunshine in England, it always seems an abiding, permanent force. The grey of yesterday, and the patter-song of the rain on the leaves, were only a dream. You were sleeping under the changeless blue of a summer night, and had but a vision of weeping, drab skies, gone now with the joy that comes in the morning. And to-morrow, when perhaps the old wild scurry of storm-cloud is alive overhead, and all the house resounds with the runnel-music from the pouring eaves, still it will be only a dream. Of a surety you will tell yourself so, as the sun breaks through the griddle of cloud, and the wind relents, and the Dutchman can get to his tailoring; and when you are stepping out amidst the swamp and glitter and rehabilitation of life, as glad of it all as the finches and butterflies that sweep on before you down the lane. The sun shines: you know it has always shone, changeless as Time itself.

With such a faith—unfounded and therefore uncontestable—I came under the glow of one brave June morning, threading field after field of blossoming clover until I stood at the gate of the bee-garden over against the hill. With its name I had long been familiar, for in the county paper there was always the little five-line advertisement, quaintly worded, announcing honey for sale. But I had never yet seen it, nor, indeed, ever set foot in this part of the good Sussex land. So, on this brimming June morning, giving rein for once to the indolent Shank's mare of moods that is fated to carry me, I set out into the bright sloth, the joyous hastelessness, of the day; and came at length to my destination—to the bee-garden that nestles under the green Downland hills.

It was girt about with a tall hedge of hawthorn, smothered in snowlike blossom, with just that rosy tinge upon it which is the first hectic of decay. Beyond the hedge I could see, stretching aloft, green apple-boughs, whose full-blown posies were alive with the desperate humming energy of countless bees. There was a blue wisp of smoke trailing idly away from a chimney-stack, all that could be seen of the snug thatched cottage within; and there were voices, a leisurely baritone, a sudden peal of laughter high-pitched and obviously a woman's, and now and then a bar or two of an old song sung in an intermittent, absent-minded way.

In one of the pauses of this song, I raised the latch of the gate. Its sharp click drew to its full lean height a figure at the end of the garden, which was bending down in the midst of a wilderness of hives. As the man came towards me coatless, his rolled-up shirt-sleeves baring wiry brown arms to the hot June sun, I took in all the busy, quiet picture. The red-tiled, winding path, the sea of old-fashioned garden-flowers on every hand, billows of lilac and red-may and laburnum, shadowy blue deeps of forget-me-not, scarlet tulips amidst them like lighthouses, and drifting shallows of amber mignonette. A decent house stood hard by, its windows bright and clean as diamond-facets. There was a gay flicker of linen on a line beyond. An old dog lolled in a straw-filled barrel. A cat kept company with a milk-jug on the spotless doorstep. And everywhere there were bee-hives, each of a different harmonious shade of colour, not ranged in stilted rows, but scattered here and there in twos and threes in the orderless order beloved of bees and unsuburban men.

The bee-master had keen grey eyes, set deep in a sun-blackened, honest face, and the ever-ready tongue of him was that of the beeman all the world over. He was ripe and willing to talk of his work, explaining what he was, and what he had done, as we slowly wandered through his domain. He was a Londoner—he told me—at least, that was his fate half a dozen years ago—a City clerk, pale as the ledger-leaves that fluttered through his fingers from nine to six of the working day. And at home, in a dreary desert of housetops called Nunhead—whither may an unkind fate never lure me—his sisters sewed for a living, white-faced as himself. But one day, in an old second-hand book-shop, he lit upon a threepenny treasure—a book on the management of bees. He read it as his train crawled homeward on one stifling, freezing, fog-bound winter's night; and there and then, in the mean, dirty cattle-box of a third-class carriage, in fancy the bee-garden was inaugurated, that has since developed into all I saw around me on that brave morning in June.

It was a long time in the doing, he told me, as we sauntered among the busy hives, speaking with a delightful Sussex intonation already veneered upon his Cockney brogue—a long and weary and scraping time. There was money to be saved, the capital needed for the enterprise; and this was no easy matter out of a total family income of forty shillings a week. But at last it was done, and well done. There came a day when the three of them shook the dust of Nunhead from their feet, and took over possession of the little tumble-down cottage with its bare half-acre of neglected ground. Well, those were hard times to begin with—he said, with an unaccountable relish in the recollection;—but now, look how all was changed! He waved a triumphant, proudly proprietary arm around him. The cottage was sound and well furnished throughout. The three or four bought hives, with which he had started his business, had multiplied into sixty or seventy, all made by his own hands. Where had he got the bees? Well, that threepenny book had taught him a secret—the art of bee-driving. Nearly all the cottagers for miles round were in the habit of sulphuring their bees to get at the honey. The first autumn, and every autumn since then, he had gone to his neighbours and told them he would take the bees out of the hives for them, and leave them all the combs and a good trinkgeld into the bargain, if they would let him have the bees for his trouble. And they were more than willing. And thus he had gradually built up his little principality of hives.

But, the profit of the thing? This, indeed, was nothing much to boast of. He sold all the honey and wax he got, sending it away, for the most part, by post, and extending the circle of his custom by little and little with every year. Taking the bad years with the good, he had made a net return of £2 for every hive; in bumper-seasons it was always much more. It was not a great deal, but there were only three of them, and their wants were simple. Their greatest needs—fresh air, peace, and quiet, the healthful life of the country—these were to be had for nothing at all. And as for clothes—you never know, until you give over trying to keep up appearances, how very little appearances count in the world. At any rate, for them, the whole thing was a complete success. There were men round about that country-side who farmed whole provinces, and still grumbled; but here was he, getting peace and plenty from half an acre; and as for the girls, they did nothing but laugh and sing all day long.

Thus we wandered and talked; and I—feigning ignorance of bee-matters, lest he might think I was but carrying coals to Newcastle in clumsy charity—bought honey, and asked many questions; and slowly the entire meaning of what had been done by these emancipated slaves of City clerkdom was revealed. The bee-master pushed his old straw hat back over his clever forehead, and lit the most comfortable pipe I had ever set eyes on. He had evidently thought the whole thing out long ago, and got it down to its essential elements.

"What we are doing here," he said, "could be done by hundreds of others who are still in London in what was once our old plight. Large bee-farms are all very well, but they are more or less a thing of the future—something that is still to be evolved out of twentieth-century needs. But the bee-garden has its immediate use and place in every district where there is an average population. People generally have got out of the habit of eating honey because it is so seldom on sale in the shops; but if you steadily and continuously remind them of it, they will buy, and soon grow to wonder how they did without it for so long. But it must be set before them in an attractive way. Run-honey must be bright and pure to look at, and neatly bottled and labelled. If you sell honey in the comb, the section-boxes must be spotlessly clean and white. In that old book that first led me to bee-keeping, it says that only the English bee should be kept, because it is a better honey-gatherer. But, from the salesman's point of view, there is a much more weighty reason for abjuring all foreign strains of bees. English bees leave a thin film of air between the honey and the cell cappings, and the result is that the comb always looks perfectly white. But nearly all foreigners fill their cells to the brim, and this means that the finest honeycomb will have a dark and dirty appearance, and no one will be tempted to buy. That is the sort of thing a business-man thinks of first, so the old training days in London have not been altogether without their use even here."

A Forest Apiary

The song, aloof and desultory, that I had heard from the garden-gate, was growing clearer as we walked; and now we turned the house-corner, and came upon more hives, with a neat, girlish figure busy among them; and, hard by, a tiny laundry-shed, wherein I caught a glimpse of brown arms deep in a wash-tub, and heard the last stanza of the vagulous song.

"Hetty, there," explained the bee-master, "helps in the garden, and— Helps, did I say? Why, she is far and away a better hand at it than I. There is so much in hive-work that needs the light touch which only a woman can give. And Deborah, she keeps house for us. Did you know that the word Deborah was Hebrew for a honey-bee? But come and see where I make the hives on winter days, and where we sling the honey, and fill the super-crates with the sections, and all the rest of it."

He showed me then his workshop and a little gauze-windowed shed where there was a homemade honey-extractor—a cunning, centrifugal thing by which the combs could be emptied and restored unbroken to the bees, to be charged again and again. And there was a storehouse, where long rows of honey-jars, and stacks of sections, and blocks of pale yellow wax were waiting for the purchaser, and a packing-shed where the post-boxes of corrugated cardboard were made up. Finally there was pointed out to me, in a far-off corner of the garden, a donkey—shaggy, well-fed, placidly browsing—and, under a neighbouring pent-roof, a little cart that was a curiosity in its way. Its wooden tilt was made to represent a big bee-hive, and on it was painted the name of the bee-garden and a list of hive-products which it carried for sale. The bee-master put an admiring hand upon it.

"It was all Hetty's idea," he said. "London girls for pluck, you know! And she goes into the town with it once a fortnight in the season; takes it away crammed full, mind, and never brings back an ounce! Somehow or other, I think those girls ought to change names!"

Journeying back to the railroad-station under the eternal English sunshine and through the chain of blossoming fields, I listened to the chant of the bees around me; and though it was the familiar sound of a lifetime, there was something in it then which I had never heard before. The rich note rose and fell; died down to silence as the path led through impregnable red-clover; swelled again as the land paled to the rosy hue of the sainfoin; burst out into a loud, glad symphony where a patch of charlock blent its despised, uncoveted gold with the farmer's drill. "You thought you knew our ways of life from Alpha to Omega"—so seemed to run, in fancy, the wavering refrain. "You have pried upon us day and night, in season and out of season. You have chloroformed us, vivisected us, torn our dead sisters limb from limb to feed the cruel, glittering eyes of that binocular of yours. You have come at last to think that there was nothing about us, within or without or round about, that you had not got to know. And here a common City clerk, turned tail on his hereditary duty, has shown you, in one short hour, a whole sheaf of things about us which you—Peeping Tom that you are!—in a whole life's keyhole-prying have never guessed. Out upon you! You deserve to have to do with nothing better than bumble-bees for the rest of your days!"

For the more I thought of little bee-gardens, such as the one I had just visited, established here, there, and everywhere throughout the land, the plainer it became that this, after all, was a mission for the honey-bee that had quite escaped me; and the fonder of the idea I grew. With bee-keeping on a grand scale there was the difficulty that an apiary might become too large for the resources of the country about it, although it is all but certain that crops grown specially for bees can be made to pay. But a small garden could never exhaust the land within its necessary three-mile radius, and all the nectar its bees could gather would be obtained free. Nunhead has done it gloriously, thought I, tramping steadily onward through the clover. And why not all the other Nunheads that hem in the great cities? There must be plenty who love the dust and din, and are willing to stop there; so the little band of bee-gardeners will never be missed.

And there was something else I thought of, too, as I strode along under the English sunshine which lasts for ever, swinging my box of superfluous, yet much-prized honey as I went.

The song and that pleasant ripple of laughter—they were in my ears still, and mingling with the labour-song of the wayside bees. Now, only a dozen miles or so, away over the hill-tops in the blue Sussex weald, I knew of just such another bee-garden, where two brothers—not Londoners this time, but true-born Downland lads—had well established themselves, were getting comfortably off, but were still single men. And only a week ago they had deplored this fact to me, and— But avast! Match-making was never yet to be reckoned part of the Lore of the Honey-Bee.